Earth For All Advance Praise

It's time to shift from the age of endless growth to an age of thriving in balance. This thought-provoking analysis proposes five critical turn-arounds for getting us there—each of them raising urgent issues for public discussion and action. Read on to explore possible futures for humanity and join the most vital debate of our times.

—Kate Raworth, author, *Doughnut Economics*

This tremendous collaboration, documented in this breathtaking book, provides yet more evidence that so much good can come out of pooling our minds and skills, to build a world that works for all. Why not opt for one-planet prosperity, if the alternative is one-planet misery?

—Mathis Wackernagel, Ph.D., founder, Global Footprint Network, co-author, *Ecological Footprint*

If we'd paid attention to *The Limits to Growth* in 1972, we wouldn't be in the fix we're in today; as the modeling in this book makes clear, what's left of this decade may be our last best hope to get it at least partly right.

—Bill McKibben, author, *The End of Nature*

This latest, most urgent, and most carefully researched version of system science's scenarios for our human future is essential reading for collapse preventers everywhere. Whether its recommendations are taken up by policy makers everywhere—and whether we humans are therefore able to avert worldwide ecological, economic, and social breakdown sometime during the remainder of the 21st century—is up to all of us.

—Richard Heinberg, senior fellow, Post Carbon Institute, author, *Power: Limits and Prospects for Human Survival*

An extraordinary book at an extraordinary time. For today and to-morrow's leaders, *Earth for All* is a must-read. This book offers a con-crete, breakthrough vision on how to ensure well-being for all—in any country—on our finite planet. Together, we can build a world that is genuinely equitable by following the 5 Turnarounds—a roadmap to accelerate the realization of the Sustainable Development Goals in the next decade. I hope it will inspire a new movement of minds and souls that are willing to save our precious humanity.

—Ban Ki-moon, 8th Secretary General of the United Nations, and Deputy Chair of The Elders

Earth For All conclusively shows that humanity's future on a livable planet depends on drastically reducing socio-economic inequality and a more equitable distribution of wealth and power. Essential reading on our long journey toward an "Earth for All" society.

—Thomas Piketty, author, *Capital in the Twenty-First Century* and *A Brief History of Equality*

The ideas explored in *Earth For All* should be discussed by all the parliaments of the world. We need to change our economies so that we start putting people before profit. And we need the rich and the polluters to pay their share for the loss and damage that the climate crisis is already unleashing on poor, vulnerable communities around the globe. It's well past time for us to create a world that's fair and just for all.

—Vanessa Nakate, climate activist, and founder, Africa-based Rise Up Movement

This book arrives at a moment in time when humanity is facing its most consequential decade in human history. What we do now will determine whether we have a future to protect. In order to ensure our survival, we must understand the interconnected nature of the current convergence of crises we are dealing with. *Earth for All* illustrates this understanding and uses it to show us a path forward that will put the wellbeing of people and our planet first, instead of profit and growth.

—Kumi Naidoo, global ambassador, Africans Rising for Justice, Peace, and Dignity

Human actions that defend our current economic model are increasingly destroying our planet, creating poverty, inequality and exclusion, failing to respond effectively to health risks, inflaming conflict—in short threatening our jobs, our communities, and our common security. *Earth For All*, provides a call to action to navigate this century with people and planet at the heart of shared prosperity. This is a roadmap that cannot be ignored.

—Sharan Burrow, General Secretary, International Trade Union Confederation (ITUC)

Earth for All plots the course to a sustainable wellbeing future that can overcome our ongoing societal addiction to growth. We all need it now more than ever.

—Robert Costanza, Professor of Ecological Economics, Institute for Global Prosperity (IGP), University College London (UCL)

Too many cooks, they say, but in this case we are talking chefs. Indeed, the multiple authorship of *Earth For All* ensures both hugely satisfying food for thought and high-energy fuel for action. The two scenarios, "Too Little, Too Late" and "The Giant Leap," will help readers to confront the existential realities now facing us, while the proposed recipes for change will help guide the those of us who value the future and are ready to roll up our sleeves in pursuit of better futures for all.

—John Elkington, founder and chief pollinator, Volans, and author,
Green Swans: The Coming Boom In Regenerative Capitalism

Earth for All is a playbook to catch up after 50 years of systemic inaction on mitigating the risks which were factored in *The Limits to Growth* in 1972. We don't have 50 years this time, we have at best 10 years to urgently engage in the five critical turnarounds. There is no way for you and me to become the system-change leaders that the world needs without starting from those five for our roadmap. Put them on your immediate to-do list.

—Emmanuel Faber, Member of the Earth4All
21st Century Transformational Economics Commission

Examining the multiple crises confronting the world and offering practical solutions is a very ambitious undertaking. The solutions offered here may be difficult for those benefiting from the broken system, but the truth remains that the planet has limits and inaction will be extremely expensive. It is either we act now or face uncontrollable disruptions. Leaders may argue that they cannot do all that is needed, but it will be a big shame to read this book and do nothing.

—Nnimmo Bassey, author, *To Cook a Continent:
Destructive Extraction and the Climate Crisis in Africa*

Timely, brilliant book. *Earth For All* describes a concrete roadmap to transform our economies and defeat poverty whilst protecting planet Earth. An inspirational read for both grassroots groups and national leaders. Will we hear? Will we make this change happen?

—Sheela Patel, founder and director,
Society for Promotion of Area Resource Centres (SPARC), Mumbai

Earth For All shows us how to make the turnarounds we need to meet the challenges facing the planet and the people. This is essential reading for everyone who would like to put their shoulder to the wheel and join the movement for transformational change.

—Kate Pickett, Professor of Epidemiology, University of York

50-years after *The Limits To Growth*, *Earth For All* says it is possible to build a prosperous future for all on our planet and shows how. This book is an absolute must-read for policymakers and leaders. With the hope that this decade is decisive to understand that Earth should indeed be for all.

—Janez Potočnik, former European Commissioner for Environment, 2009–2014, former Minister for European Affairs for Slovenia, co-chair, International Resource Panel (IRP)

Earth for All is an extraordinary, potentially historic, breakthrough guide to a viable and fulfilling future for all on a finite living Earth. My highest recommendation. Read it. Share it. Discuss it.

—David Korten, author, *When Corporations Rule the World*, *The Post-Corporate World: Life After Capitalism*, and *Change the Story, Change the Future: A Living Economy for a Living Earth*

I've always kept the original *The Limits to Growth* report within easy reach. Now I'll be putting *Earth for All* beside it. An essential guidebook for anyone aspiring to be a good ancestor.

—Roman Krznaric, author, *The Good Ancestor: How to Think Long Term in a Short-Term World*

For the first time we have a narrative about our future that is neither utopia nor collapse and that is endorsable across the political spectrum. It is an aspirational future. It is livable for all, and most crucially it is achievable.

—Carlota Perez, author, *Technological Revolutions and Financial Capital*

The health of humanity is increasingly threatened by multiple environmental changes, driven by inequitable and unsustainable patterns of consumption. The economic transformation described in *Earth For All* can support the achievement of health for all and provide the opportunity for societies around the world to flourish within planetary boundaries. It should be read by everyone who is concerned about the future.

—Andy Haines, Professor of Environmental Change and Public Health, London School of Hygiene and Tropical Medicine

Though curiously silent on the deep cultural and spiritual revolutions required and the pluriverse of cosmologies available for this, the five strands of the Great Leap proposed by the authors—poverty, inequality, gender, food, and energy—are crucial to the fundamental transformations we need to make peace with ourselves and the earth.

—Ashish Kothari, Kalpavriksh and Global Tapestry of Alternatives, co-editor, *Pluriverse*

Fifty years after the forward-looking publication of *The Limits to Growth*, this new report to the Club of Rome provides the most compelling and practical blueprint for socioeconomic transformation here and now, with a view to avoiding climate catastrophe and building a better society for everyone.

—Lorenzo Fioramonti, author, *The World After GDP: Economics, Politics and International Relations in the Post-Growth Era*, and member of the Italian Parliament

Earth for All clearly illustrates how the fight against inequality and poverty is a precondition to stop climate change and protect the planet. This book is a call for all governments of the world to upgrade their economic systems. A must read.

—Jane Kabobo-Mariara, president, African Society for Ecological Economics

Earth for All provides us with a vision of a future where humanity and nature are in balance and wellbeing is at the core of our economic system. But it also provides us with the actions we need to take to get there. This book is a must read for policymakers.

—Ernst von Weizsäcker, Honorary President, the Club of Rome

This now-or-never moment in history to avoid ecosystem collapse is also a profound opportunity for humanity to rediscover its purpose in human and ecological thriving. *Earth for All* tells us where and how to start.

—Gaya Herrington, Vice-President ESG Research, Schneider Electric

Threading the needle between what is good investment and what is good for society is hard. *Earth For All* provides a powerful new framework that's a must read for every impact investor.

—Doug Heske, founder, Newday Impact Investing

This is rigorously joined up thinking that answers credibly the challenge of our time. *Earth for All* charts a scientifically tested route towards an inclusive, clean, modern economy supported by thriving natural systems and a regenerative agricultural model. Should be adopted by all policy makers.

—Charles Anderson, Former Director UNEPFI, Chairman CO2eco

Earth For All sharply addresses the biggest challenge of our time: how to defeat inequality and poverty, whilst saving our planet from climate change and environmental destruction. Its urgent call to transform our economies is my call to our leaders. A must-read for all of us.

—Naoko Ishii, Professor and Executive Vice President at the University of Tokyo. Former CEO of the Global Environmental Facility (GEF)

Stubborn optimism versus immobilizing pessimism. Long-term vision versus short-term reaction. Collective intelligence versus individualism. Human well-being vs compulsive consumption. Valuing our future versus discounting it. A livable planet versus an unstable planet. The choices we must make for a prosperous common future are crystal clear. So is the urgency to act and to redress the imbalances of a broken socio-economic model. What is less clear is how to articulate the system change we need, how to manage the complexities that come with it, how to constructively engage all the relevant stakeholders, how to sequence the moves from the different players, how to prioritize strategic transformations, how to measure impacts, how to anticipate and to mitigate risks... And here is where *Earth for All* comes to the rescue: an honest contribution for positive change from some of the most renowned thinkers, scientists and economists of our time. A recalibrated set of lenses to explore the challenges of our generation: global equity and a healthy planet. A map to explore, dive deep and inspire. A must-read for any policymaker who values our future, as well as for corporate leaders, responsible investors and the general public worldwide. *Earth for All* is a call for action and a movement to infuse social and political change for the common good. *Earth for All* is inspired by the legacy of *The Limits to Growth*, but it goes well-beyond that. It provides a guide to leapfrog into the future most of us long for. This is the tale of our time. A story not to be missed.

—Teresa Ribera, Deputy Prime Minister for
the Ecological Transition, Government of Spain

Timely and important. A major contribution to a better shared future for humanity.

—Jinfeng Zhou, Secretary General of the China Biodiversity
Conservation and Green Development Foundation

Earth for All is a vision for a possible future based on global and local action. We hope it will spark honest, bold conversations, and help people around the world make decisions to redesign their societies.

—Chandran Nair, author, *Dismantling Global White Privilege:*
Equity for a Post-Western World

Earth for All

A SURVIVAL GUIDE for Humanity

A Report to the Club of Rome (2022)
Fifty Years After *The Limits to Growth* (1972)

by

Sandrine Dixson-Declève | Owen Gaffney
Jayati Ghosh | Jorgen Randers
Johan Rockström | Per Espen Stoknes

new society
PUBLISHERS

Written by Sandrine Dixson-Declève, Owen Gaffney, Jayati Ghosh,
Jorgen Randers, Johan Rockström, and Per Espen Stoknes

Cover design by Diane McIntosh.
Cover images: © iStock.

First printing LSI version September 2022.

Inquiries regarding requests to reprint all or part of *Earth for All*
should be addressed to New Society Publishers at the address below.

To order directly from the publishers, order online at www.newsociety.com

Any other inquiries can be directed by mail to:
New Society Publishers
P.O. Box 189, Gabriola Island, BC V0R 1X0, Canada (250) 247-9737

LIBRARY AND ARCHIVES CANADA CATALOGUING IN PUBLICATION

Title: Earth for all : a survival guide for humanity : a report to the Club of Rome (2022),
fifty years after The limits of growth (1972) / by Sandrine Dixson-Declève, Owen Gaffney,
Jayati Ghosh, Jørgen Randers, Johan Rockström, Per Espen Stoknes.

Names: Dixson-Declève, Sandrine, author. | Gaffney, Owen, author. | Ghosh, Jayati,
author. | Randers, Jørgen, author. | Rockström, Johan, author. | Stoknes, Per Espen,
author. | Club of Rome, addressee.

Description: Includes bibliographical references and index.

Identifiers: Canadiana (print) 20220274843 | Canadiana (ebook) 20220274851 |
ISBN 9780865719866 (softcover) | ISBN 9781550927795 (PDF) |
ISBN 9781771423755 (EPUB)

Subjects: LCSH: Climatic changes—Forecasting. | LCSH: Climate change mitigation—
Social aspects. | LCSH: Climate change mitigation—Economic aspects. | LCSH:
Climate change mitigation—Political aspects. | LCSH: Environmental policy. |
LCSH: Environmental protection—Social aspects. | LCSH: Environmental protection—
Economic aspects. | LCSH: Environmental protection—Political aspects.

Classification: LCC QC903 .D59 2022 | DDC 363.738/74—dc23

New Society Publishers' mission is to publish books that contribute in fundamental
ways to building an ecologically sustainable and just society, and to do so with the least
possible impact on the environment, in a manner that models this vision.

Contents

Contributors

MAIN AUTHORS

Sandrine Dixson-Declève, Owen Gaffney, Jayati Ghosh, Jorgen Randers, Johan Rockström, Per Espen Stoknes

CONTRIBUTING AUTHORS

TEC= Members of the 21st Century Transformational Economics Commission:
Anders Wijkman (TEC), Hunter Lovins (TEC),
Dr. Mamphela Ramphele (TEC), Ken Webster (TEC)

CONTRIBUTORS

Nafeez Ahmed (TEC), Lewis Akenji (TEC), Sharan Burrow (TEC), Robert Costanza (TEC), David Collste, Emmanuel Faber (TEC), Lorenzo Fioramonti (TEC), Eduardo Gudynas (TEC), Andrew Haines (TEC), Gaya Herrington (TEC), Garry Jacobs (TEC), Till Kellerhoff, Karthik Manickam, Anwesh Mukhopadhyay, Jane Kabubo-Mariara (TEC), David Korten (TEC), Nigel Lake, Masse Lo, Chandran Nair (TEC), Carlota Perez (TEC), Kate Pickett (TEC), Janez Potočnik (TEC), Otto Scharmer (TEC), Stewart Wallis (TEC), Ernst von Weizsäcker (TEC), Richard Wilkinson (TEC)

DATA SYNTHESIS, SYSTEM ANALYSIS, AND MODELLING TEAM

Jorgen Randers, Ulrich Golüke, David Collste, Sarah Mashhadi, Sarah Cornell, Per Espen Stoknes, Jonathan Donges, Dieter Gerten, Jannes Breier, Luana Schwarz, Ben Callegari, Johan Rockström

SUPPORTING DEEP DIVE PAPERS (AVAILABLE AT WWW.EARTH4ALL.LIFE)

Nafeez Ahmed, Shouvik Chakraborty, Anuar Sucar Diaz Ceballos, Debamanyu Das, Jayati Ghosh, Gaya Herrington, Adrina Ibnat Jamilee Adiba, Nigel Lake, Masse Lô, Chandran Nair, Rebecca Nohl, Sanna O'Connor, Julia Okatz, Kate Pickett, Janez Potočnik,

Dr. Mamphela Ramphele, Otto Scharmer, Anders Wijkman, Richard Wilkinson, Jorgen Randers, Ken Webster

EDITORS
Joni Praded, Ken Webster, Owen Gaffney, and Per Espen Stoknes

EARTH4ALL PROJECT MANAGEMENT AND SUPPORT
Per Espen Stoknes (Scientific Work Packages), Sandrine Dixson-Declève, Anders Wijkman (TEC), Owen Gaffney (Communications), Till Kellerhoff (Coordination)

EARTH4ALL CAMPAIGN TEAM AND BOOK STORY DEVELOPMENT
Philippa Baumgartner, Rachel Bloodworth, Liz Callegari, Lena Belly-Le Guilloux, Andrew Higham, Nigel Lake, Luca Miggiano, Zoe Tcholak-Antitch

ACKNOWLEDGMENTS
Azeem Azhar, Tomas Björkman, Alvaro Cedeño Molinari, John Fullerton, Enrico Giovannini, Maja Göpel, Steve Keen, Connie Hedegaard, Sunita Narain, Julian Popov, Kate Raworth, Tom Cummings, Petra Künkel, Grace Eddy, Megan McGill, Roberta Benedetti, Vaclav Smil, Julia Kim, Roman Krznaric, Sir Lord Nicholas Stern, Andrea Athanas, Kaddu Sebunya

FUNDERS
Angela Wright Bennett Foundation, Global Challenges Foundation, Laudes Foundation, Partners for a New Economy

GRAPHICS
Les Copland, Philippa Baumgartner

MEMBERS OF THE 21ST CENTURY TRANSFORMATIONAL ECONOMICS COMMISSION
Nafeez Ahmed, Director of Global Research Communications, RethinkX; and Research Fellow, Schumacher Institute for Sustainable Systems
Lewis Akenji, Managing Director, Hot or Cool Institute
Azeem Azhar, Founder, Exponential View
Tomas Björkman, Founder, Ekskäret Foundation

Sharan Burrow, General Secretary, International Trade Union Confederation (ITUC)

Alvaro Cedeño Molinari, Former Costa Rican Ambassador to Japan and the WTO

Robert Costanza, Professor of Ecological Economics, Institute for Global Prosperity (IGP) at University College London (UCL)

Sandrine Dixson-Declève, Co-President, The Club of Rome and Project Lead, Earth4All

Emmanuel Faber, Chair, International Sustainability Standards Board

Lorenzo Fioramonti, Professor of Political Economy, and Member of the Italian Parliament

John Fullerton, Founder and President, Capital Institute

Jayati Ghosh, Professor of Economics, University of Massachusetts Amherst, USA; formerly at Jawaharlal Nehru University, New Delhi

Maja Göpel, Political economist and transformation researcher

Eduardo Gudynas, Senior Researcher, Latin American Center on Social Ecology (CLAES)

Andy Haines, Professor of Environmental Change and Public Health, London School of Hygiene and Tropical Medicine

Connie Hedegaard, Chair, OECD's Roundtable for Sustainable Development, former European Commissioner

Gaya Herrington, Vice-President ESG Research at Schneider Electric

Tim Jackson, Professor of Sustainable Development and Director of CUSP, the Centre for the Understanding of Sustainable Prosperity at the University of Surrey

Garry Jacobs, President & CEO, World Academy of Art & Science.

Jane Kabubo-Mariara, President of the African Society for Ecological Economists, : ED, Partnership for Economic Policy

Steve Keen, Honorary Professor at University College London and ISRS Distinguished Research Fellow

Julia Kim, Program Director, Gross National Happiness Centre, Bhutan

Roman Krznaric, Public philosopher and author

David Korten, Author, speaker, engaged citizen, and president of the Living Economies Forum

Hunter Lovins, President, Natural Capital Solutions; Managing Partner, NOW Partners

Chandran Nair, Founder and CEO, The Global Institute for Tomorrow

Sunita Narain, Director-General Centre for Science and Environment, Delhi and editor, Down To Earth

Carlota Perez, Honorary Professor at IIPP, University College London (UCL); SPRU, University of Sussex and Taltech, Estonia.

Janez Potočnik, Co-chair of the UN International Resource Panel, former European Commissioner

Kate Pickett, Professor of Epidemiology, University of York

Mamphela Ramphele, Co-President, The Club of Rome

Kate Raworth, Renegade economist, creator of the Doughnut of social and planetary boundaries, and co-founder of Doughnut Economics Action Lab.

Jorgen Randers, Professor Emeritus of Climate Strategy, BI Norwegian Business School

Johan Rockström, Director of the Potsdam Institute for Climate Impact Research

Otto Scharmer, Senior Lecturer, MIT, and Founding Chair, Presencing Institute

Ernst von Weizsäcker, Honorary President, The Club of Rome

Stewart Wallis, Executive Chair, Wellbeing Economy Alliance

Ken Webster, Director International Society for Circular Economy

Anders Wijkman, Chair of the Governing Board, Climate-KIC, Honorary President, The Club of Rome

Foreword

by Christiana Figueres

Former Executive Secretary of the United Nations Framework Convention on Climate Change (UNFCCC), one of the architects of the Paris Agreement, co-founder of Global Optimism, and co-host of the climate podcast *Outrage + Optimism*.

Millions of people around the world are suffering deeply as a result of climate chaos, environmental degradation, and perverse inequality. For way too long, the multilateral system and civil society have defined and described those multiple crises as separate, each with their own unique set of solutions, often in competition with each other. In fact, they are different aspects of what we might understand as the metacrisis.

Earth for All shows how we address these crises together, and that's what makes it such critical reading. This is a path of possibility, infused with stubborn, urgent optimism. *Earth for All* does not gloss over the facts or current context, nor does it offer a utopian vision for the future. What this book shows us is that it *is* possible to avoid rising social tensions, rising human suffering, and rising environmental destruction by making five extraordinary turnarounds in the interconnected challenges.

As we prepare for these challenges, it would serve us well to understand that they are interconnected not only in their social and economic realities but more fundamentally in their source. The climate crisis, the nature crisis, the inequality crisis, the food crisis all share the same deep root: extractivism based on extrinsic principles. This extractivism does not only deplete the planet—the very soil of the Earth itself—it also depletes our human souls.

In order to take forward the good and necessary work to regenerate our planet and societies, to turn around our economic systems so that we can see the positive changes with our own eyes, we must also regenerate what is internally palpable to each of us.

We are going to need a nourishing and optimistic mindset to muster the courage necessary to transform economies so that human and planetary wellbeing come first. The economy after all is a system that we humans designed. In its current form, the global economy reflects a chronic neglect of our inner world and what human beings hold most dear. We reward competition instead of cooperation. We reward environmental destruction instead of balance with nature. We reward short-term gains instead of long-term peace and prosperity for future generations.

To turn this around, the invisible, internal world within each of us needs regenerating too, with compassion and solidarity for ourselves and each other. The metacrisis is not only extrinsic—exiting outside of ourselves, it is intrinsic—existing within ourselves.

When I took the role of Executive Secretary at the UNFCCC, I was asked during my first press conference whether I thought a global agreement on climate change was possible. I blurted out: "Not in my lifetime!" But as soon as I said those words, which accurately reflected the prevailing mood, I realized that if we were to achieve a global deal, I was going to have to change my attitude. I, personally, would have to become a beacon of possibility. So I set about the deep work to transform people's attitudes, starting with myself. The journey was long, difficult, and involved thousands of people working together. The end result was the historic Paris Agreement on climate change just a few years later.

Large-scale systems change is surprisingly personal. It starts with each of us, with what we prioritize, what we are willing to stand up for, and how we decide to show up in the world: we are the authors of the next chapter of humanity.

So, I encourage the reader—especially if you are a leader in your community, company, or city—to pause for a moment before you dive into this most excellent text to turn around and face yourself. Think very carefully and intentionally about the Giant Leap you must—and can—make inside yourself in order to contribute fully to the extraordinary Giant Leap for which *Earth for All* so generously offers us the road map.

Foreword

by Elizabeth Wathuti

Climate activist and founder, Green Generation Initiative

Some of the best and most powerful moments in my life have been sitting by the side of the river, watching the water flowing, and seeing the trees swaying in the wind. In those moments, it's possible to feel our true connection to nature. Nature is the air we breathe, the food we eat, and it's connected to our health and wellbeing. The beauty of nature can make us feel happy; it's how we can feel at peace.

Seeing nature destroyed can make us feel very angry; it makes me feel angry. Even as I work to help children plant trees within their school compounds every day, around the world, enormous machines still cut down entire forests faster than we can snap our fingers, extracting "wealth" for export. The rivers are polluted with toxic chemicals and plastic that destroy our ability to find joy by sitting on the riverbanks, and to access clean drinking water.

The humanitarian crisis resulting from this ecocidal relationship with the Earth is rapidly worsening. Poverty and inequality are creating unbearable differences between and within countries. Today, millions of people across the Horn of Africa are facing climate-related starvation. The situation is shocking. I have seen lives and livelihoods devastated by drought in my home country, Kenya. I have visited and talked to communities in Wajir who are losing hope for the future.

And, even as they lose their livestock and suffer greatly, many in these rural communities do not yet know the scale of the climate crisis. They do not know that the crisis they are experiencing is happening all over the world, across boundaries, as a result of a global economic system that is horribly broken.

At the same time, it seems that our world leaders do not really understand or feel the pain of the climate crisis and its devastating impact on people, despite what they have been told. They seem not to see that the system we have now is not working for most people.

Professor Wangari Mathai, one of my greatest inspirations, said "Those of us who understand, who feel strongly, must not tire. We must persist. The burden is on those who know. We are the ones who get disturbed and are caused to take action."

I have asked our world leaders to open their hearts and feel the pain and suffering. I have asked them to listen to the truth and act with compassion because I believe that the will to act must come from deep within. We have a human capacity to care deeply and to then act.

If we open our hearts, the seeds of transformative action will flourish. We *can* take a Giant Leap from the interconnected crises we face now into a future with a stable climate, clean air, clean water, and food security for all. But to do so, we need to change our way of thinking, and we need to start telling new stories about what is important and what is possible. That's why the stories in this book are so important, because they put people's wellbeing at the heart of the solutions.

Eliminating poverty and addressing inequality together, in order for our societies to be able to address the climate crisis and its impact, is exactly what Wangari Mathai stood for when she began her courageous work planting trees. At the heart of Professor Mathai's work to stop deforestation was her goal to empower women: to give them the means to provide fuel, food, shelter, and income to support the education of their children.

Earth for All is a call to participate in this interconnected work and thinking. It's a reminder of how enormous the transformations we must make are, that it is possible to change the system we have built; and it's full of new ideas about how to put wellbeing and dignity, and cooperation and solidarity, right at the root of action.

I love the idea of citizens assemblies as a powerful way to bring the voices of the people and their ideas for change to our global leaders, and I hope thousands of these assemblies can bloom. I know from Professor Mathai's legacy and my own work that people can be agents of their own future, even when the challenges they face are overwhelming. And I know that our leaders do have hearts. My wish is that they open their hearts, so that we can all work together for a future that this book says can be achieved.

I hope you can find a beautiful tree or river to sit by and read it.

Earth for All

Five Extraordinary Turnarounds for Global Equity on a Healthy Planet

This is a book about our future—the collective future of humanity this century, to be precise. Civilization is at a unique moment, a juncture. Pandemics, wildfires, and wars swirl around us as we write, sure signs that societies remain extremely vulnerable to shocks despite unprecedented progress. Beyond the immediate turbulence, we are in the midst of a planetary emergency of our own making. What this book will argue is that the long-term potential of humanity depends upon civilization—a wondrous, freewheeling, kaleidoscopic, inspiring, confounding civilization—undergoing nothing short of five extraordinary turnarounds within the coming decades.

We know the pain points. Everyone knows we must end extreme poverty for billions. Everyone knows we must fix the inequality crisis. Everyone knows we need an energy revolution. Everyone knows our industrial diets are killing us, and the way we farm food is ripping through nature, driving a sixth mass extinction of species. We know human populations cannot increase endlessly. And we know our material footprint cannot expand infinitely on our finite, blue and green Earth.

Can "we"—meaning all people and peoples—come together to navigate this century? Can we take a collective leap in human development with courage and conviction? Can we overcome divisions, neocolonial and financial exploitation, historic inequalities, and deep, deep distrust among nations to deal with the long-term emergency? Can we achieve *systemic transformation in decades, not centuries?*

Our goal with *Earth for All* is to show you that this is indeed fully possible. And that it won't cost the Earth. Rather, it is an investment in our future. Based on expert assessments supported by system dynamics models, the pages ahead explore the most likely routes to emerge from these emergencies; the pathways that bring the most humanitarian, social, environmental, and economic benefits to all.

Earth for All is about valuing our future. Most people value their personal future. But what about valuing our collective future? As a civilization, as eight billion people, as an entangled web of societies? Well, the evidence that we do is very limited. The COVID-19 pandemic is certainly a prime example of this failure. Despite enormous wealth in some countries, we simply did not put in place basic safeguards to protect civilization from a threat that was known, highly likely, and entirely avoidable. The investment in adequate preparation was, essentially, peanuts compared with the global suffering to date.

Another sign of chronic failure: Millions of children have had to walk out of schools around the globe and march in the streets to get our attention. The school strikers' message is simple: "Our house is on fire." Those with power, they say, are taking colossal risks with their future, consigning them to live on a destabilized Earth. The placards on the streets read "Systems Change, Not Climate Change" and "Listen to the Science." And the youth carrying them are demanding, rightly, a fair and just transition of societies. Now.

Their plea exposes some uncomfortable questions. Why are actions to prevent pandemics or climate disruptions so shockingly inadequate? Are economic systems driving industrial societies in a direction that's impossible to change? Is it even possible for everyone, whether eight or ten billion people on Earth, to prosper within planetary boundaries? Is societal collapse inevitable? Or can we find a way to value and invest in our collective future here on Earth?

This book tackles that last question head on. It presents the findings of the Earth for All initiative, which began in 2020. As the pandemic ripped through societies, an international team of scientists, economists, and multidisciplinary experts joined together to analyze what is necessary to build a fairer, more resilient economic system to weather current interconnected crises and future storms. We debated. We frequently disagreed. Some of our disagreements spilled over into heated arguments. Even with heartfelt commitment to end poverty and neocolonialism and address inequality in all societies, the perspectives between academics and authors in Europe and North America and those from Asia and Africa turn out to be quite different. For example, even though there is full agreement that a food system turnaround is essential, it was tricky navigating how much emphasis to place on organic farming, lab-based alternatives to meats, and the role of man-made chemicals during the necessary transition.

Our analysis focused on two deeply intertwined systems: people and planet, or more explicitly the global economy and Earth's life support system. It is grounded in systems thinking, a branch of science that has exploded in the last five decades and whose tools help us understand complexity, feedback loops, and exponential impacts. Systems thinkers are always on the lookout for leverage points where a small change in one thing can make a big difference to the whole system.

At the heart of the analysis are two intellectual engines that allowed us to explore the boldest economic proposals: the Transformational Economics Commission—an international group of leading economic thinkers—and a system dynamics model we call Earth4All. Through a series of feedback loops, economic ideas from the commission could be tested by the Earth4All model to see if the proposals would have a big enough effect on people and planet over time. Likewise, the commission could critique and challenge the outputs from the Earth4All model.

All of this gave us a robust process to study possible alternative future worlds. We could explore what may happen this century given a wide set of assumptions about human behavior, future technological development, economic growth, and food production—and how all of this affects the biosphere and climate. We got a glimpse of what could happen if the gap between rich and poor widens or shrinks, if greenhouse gas emissions rise or fall, if population explodes or drops, if material consumption mushrooms or is reined in, or if investment in public infrastructure and technological innovations can prevent catastrophe. While analyzing various future scenarios, the role of the model was primarily to keep our thinking straight. It helped ensure that our scenarios were internally consistent and actually followed from the assumptions we made.

Two novelties included in the model are the Social Tension Index and the Average Wellbeing Index. These allowed us to estimate whether policies—for example, related to income redistribution—might cause social tensions in societies to rise or fall. We believe that if social tensions rise too far, societies may enter a vicious cycle where declining trust causes political destabilization, economies stagnate, and wellbeing declines. In that situation, governments will struggle to deal with rolling shocks let alone long-term existential challenges like pandemic risk, climate change, or ecological collapse.

The Earth4All model operates at a global scale, which is useful for exploring big-picture long-term trends. But this can mask important regional differences. For example, global trends showing strong economic growth may hide economic stagnation in some areas. With this in mind, we developed the model further to track ten regions of the world.[1] This allows us a glimpse of how our scenarios play out in low-income countries of sub-Saharan Africa and South Asia compared with high-income countries of Europe and the United States. Of course, with any additional complexity in any model, this creates additional uncertainties so we interpret results cautiously.

Breakdown or Breakthrough?

Of all scenarios we could describe in some detail, in this book we have chosen two, which we call *Too Little Too Late* and *Giant Leap*. Too Little Too Late asks, What if the economic system driving the world (and now the biosphere) continues operating largely as it has done over the past fifty years? Will current trends in reducing poverty, rapid technological innovation, and energy transformation be enough to avoid societal collapses or Earth system shocks? Giant Leap asks, What if the economic system is transformed through extraordinary efforts to build a more resilient civilization? It explores what it may take to eliminate poverty, create trust, and provide a stable global economic system that delivers higher wellbeing to the majority. Our two scenarios are built from expert assessment and the existing academic literature, and are kept internally consistent by the Earth4All model. When we combine these, we arrive at the following conclusions.

First, on current political and economic paths, we expect continuing rising inequality by design. We also expect slow economic development in low-income countries, causing enduring poverty. As a result of inequalities within countries, social tensions are likely to rise toward the middle of the twenty-first century.

Second, these factors are likely to contribute to inadequate responses to the climate and ecological emergency. Global average temperature is likely to significantly exceed 2°C, the limit stipulated in the Paris Agreement on Climate, and established by science as a red line it would be deeply unwise to cross.[2] Large populations will increasingly face extreme heat waves, megadroughts leading to frequent crop failures, torrential rain, and rising sea levels. The world risks regional societal instabilities as a result of rising social tensions this century with global impacts. Significant parts of the Earth system are more likely than today to cross more irreversible or abrupt tipping points. This is likely to further exacerbate social tensions and conflicts. The impacts of crossing climate and ecological tipping points are likely to last centuries to millennia.

Third, five extraordinary turnarounds are needed to substantially reduce risks:

1. ending **poverty**
2. addressing gross **inequality**
3. **empowering** women
4. making our **food** system healthy for people and ecosystems
5. transitioning to clean **energy**

These extraordinary turnarounds are designed as policy road maps that will work for the majority of people. They are not an attempt to create some impossible-to-reach utopia; instead, they are an essential foundation for a resilient civilization under extraordinary pressure. And, what's more, there is sufficient knowledge, funds, and technologies in the world to implement them. These five turnarounds are not particularly new. The various actions that drive them have been described separately in many reports. But what we have attempted through Earth4All is to connect them up in one dynamic system, to assess if *together* they create sufficient economic momentum to push the global economy off the destructive course it is on and onto a resilient path.

We cannot claim this is the precise blueprint for a safe, just future. But we do claim *nothing less than focused, large-scale investment* in these five areas, starting now, is necessary. Why? Well, "just" addressing the climate emergency requires reconfiguring the global energy system—the foundation of all economies—in a single generation. Many of the engineering solutions such as solar panels, wind turbines, batteries, and electric vehicles are here already and scaling exponentially. But the solutions must be acceptable, fair, and affordable to the global middle classes or risk deep resistance. If the energy transformation already underway perpetuates historic injustices, it will have a destabilizing effect on societies. The Earth for All turnarounds show how, with a systemic approach, success might be achieved.

This brings us to the fourth conclusion. The extra investment needed to build a more resilient civilization is likely to be small: in the order of 2% to 4% of global income per year for sustainable energy security and food security.[3] But this investment is highly unlikely to emerge through market forces alone. These extraordinary turnarounds require reshaping of markets and long-term thinking. Only governments, supported by citizens, can provide this. So, the clear conclusion is that governments need to become much more active. The investments will be highest during the first decades after implementation starts, and then decline.

The fifth conclusion: Income redistribution is not negotiable. Long-term economic inequality combined with short-term economic crises (this is the current modus operandi of most large economies) contributes to economic anxiety, distrust, and political dysfunction. These are important risk factors for destructive polarization in democratic societies, which leads to rising social tensions. Because the current dominant economic model will lead to greater income inequality, extraordinary interventions are needed to address that inequality so that we can respond to global existential threats.

We propose a series of policies to ensure the wealthiest 10% take no more than 40% of national incomes. This is far from full income equality in some impossible utopia, but we estimate this is a minimum for functional democratic societies. When gross inequality corrodes trust, it becomes more difficult for democratic societies to make collective, long-term decisions that cut emissions, safeguard forests, protect freshwater, and stabilize global temperature at what scientists estimate is a relatively safe level (1.5°C). Failing to do this will in turn commit the world to even more extreme heat waves, crop failures, and food price shocks. It will worsen inequalities, erode trust further, and test governability to the limit.

Sixth, these extraordinary turnarounds can be achieved by 2050, within a single generation. But action needs to start now. Our future will be vastly more peaceful, more prosperous, and more secure if we do everything in our power to stabilize Earth this decade than if we do not. Without urgent action, we can expect rising social tensions that will make it more difficult to solve civilizational challenges in future.

Seventh, these extraordinary turnarounds will be disruptive. There is no getting away from it. The turnarounds will interact with ongoing disruptive trends, for example the next phase of the exponential technological breakthroughs. Exponential technology promises revolutions in artificial intelligence, robotics, connectivity, and biotechnology bringing economic, health, and wellbeing benefits but with massive implications for privacy, security, and the future of employment. We need to establish social safety nets during this transformation to protect all in society. This is why we have proposed Citizens Funds to distribute a "universal basic dividend" as a keystone policy innovation to address inequality and protect populations from inevitable economic disruptions. Like a traditional "fee and dividend" policy, a Citizens Fund has two parts: The private sector is charged for extracting and using resources that should be seen as under the stewardship of all in society, including fossil fuels, land, freshwater, the ocean, minerals, the atmosphere, and even data and knowledge. The fees are put into national Citizens Funds, and this revenue is then distributed back to all citizens in a country equally through a universal basic dividend (UBD).

And our final conclusion is that, despite these warnings, it is possible, desirable, and even essential to be *optimistic* about our collective future. Our analysis indicates it's fully doable. The window is still open to achieve the Earth for All vision: human wellbeing within planetary boundaries. A concerted effort to redistribute wealth can build trust within nations and between nations, opening up the space to make long-term decisions to reduce risk of existential challenges like climate disruptions or future pandemics. Rapid economic development following these five extraordinary turnarounds could remove absolute poverty by 2050. A rapid shift away from today's fossil fuels and wasteful food chains has the potential to bring long-term energy and food security to all societies. Millions of people currently enduring horrific air pollution in overcrowded cities will be able to breathe clean air again as economies transform. And a clean energy revolution driven by exponential technologies and systemic efficiencies can enable low-income countries to satisfy material needs while avoiding the historic mistakes of the rich nations. Through these extraordinary turnarounds we value our future.

The analysis clearly shows the next decade must see the fastest economic transformation in history. The scale of that transformation may seem daunting.

It is bigger than the Marshall Plan—the economic investments that rebuilt Europe after two world wars.

It is bigger than the Green Revolution in the 1950s and 60s that industrialized farming in Asia and Africa and helped eradicate famine.

It is bigger than the anticolonial movements that led to independent nations in the mid-twentieth century.

It is bigger than the civil rights movements in the 1960s that brought more equal rights to marginalized groups in the United States, Europe, and elsewhere.

It is bigger than the moon landings that cost ~2% of US gross domestic product (GDP) in the 1960s.

It is bigger than the Chinese economic miracle of the last thirty years that lifted 800 million out of poverty.

It's all of these rolled into one. On steroids. Our challenge with this book is to convince you it can be done.

It will require building the broadest coalition the world has ever seen. And it will need to happen as economic power shifts away from the old dominant West in the coming decades toward what we call in this book "Most of the World." Across regions we need an engaged majority on board: both the political left and right, the centrists and greens, nationalists and globalists, managers and workers, businesses and society, voters and politicians, teachers and students, rebels and traditionalists, grandparents and teenagers. It will require rewiring the global economic system. In particular, we need to rethink economic growth, so that economies that need to grow can grow and economies that are overconsuming can develop new operating systems.

It will require rethinking consumption of materials, which may double by 2060 without the extraordinary turnarounds.

It will require redesigning the global financial system from one that is crowdfunding catastrophe to one that crowdfunds long-term prosperity. One priority is redesigning the flow of money in the world. This means upgrading institutions like the International Monetary Fund and the World Bank to make the flow benefit those in poverty, not just the top 10%.

And, it will require more efficient, smarter, and more entrepreneurial states that look out over the horizon and put the safety of their citizens first. Governments must actively support innovation, redesign markets, and redistribute wealth.[4] So, governments need to wake up. The first duty of a state, after all, is to protect citizens from harm. In this volatile century, this means thinking in terms of systems, acting globally, and investing before it becomes profitable, in order to increase the wellbeing of future generations.

A Brief History of Future Scenarios

Earth for All builds on decades of economic and Earth system research. Let's spin the clock back fifty years. People were increasingly concerned about population growth, pollution, and the state of the planet. Rachel Carson's book *Silent Spring* published a decade earlier had triggered a real and serious fear that humans could destroy living conditions on Earth. Recognizing this, the United Nations convened the first Earth Summit—the UN Conference on the Human Environment in Stockholm. In advance of this summit, a group of young researchers based at the Massachusetts Institute of Technology (MIT) published a remarkable book, The Limits to Growth.[5]

The Limits to Growth warned about the possibility—even likelihood—of ecological overshoot and societal collapse. If humanity kept pursuing economic growth and exponential consumption without regard for finite natural resources or environmental costs, the authors warned, global society would overshoot Earth's physical limits and experience sharp declines in available food and energy along with rising pollution, an ensuing decline in standards of living, and ultimately a dramatic fall in the human population, within the first half of the twenty-first century. The book became an unlikely bestseller, with millions of copies sold worldwide.

The Limits to Growth analysis was based on a then-new computer model, World3. Computer power in the early seventies was severely limited by today's standards. But nevertheless, the MIT team created the first computer model that sought to capture the complex global dynamics of human societies evolving on a finite planet.

The team used World3 to explore future scenarios related to population growth, fertility, mortality, industrial output, food, and pollution at a grand scale. The model captured some of the complexity between, for example, the impact of population growth on food availability given that food production can't keep expanding indefinitely. Since then, other computer models have been developed to explore complex global challenges. The results presented here use the same techniques as World3. Indeed, our central model, Earth4All, was designed by Jorgen Randers, one of the four authors of The Limits to Growth.

Some of the scenarios explored in The Limits to Growth ended in collapse due to rising pollution, food production falling, and a dramatic decline in population. But not all scenarios followed this course. The team also identified a set of assumptions that produced "stabilized world" scenarios. In those, human welfare grew and remained high; key actions could be taken to avoid collapse. The media and other commentators largely ignored these scenarios, instead focusing on the threat of collapse if growth followed the traditional trajectory. Decision-makers ignored them, too, opting to remain complacent, follow neoliberal economic theory, and pursue growth at all costs— despite the warnings *The Limits to Growth* issued about the long-term effects of business as usual.

So, fifty years on, what do we make of *The Limits to Growth* scenarios? Did they align with reality in any shape or form?

What we can say with half a century of hindsight is that World3 is not just one of the most famous but also a surprisingly accurate global assessment model. In 2012, Australian physicist Graham Turner plotted real-world data from 1970 to 2000 against *The Limits to Growth* Business-as-Usual scenario. He found that the team's scenario tracked closely with reality. An updated look in 2014 showed the same.[6]

In 2021, Dutch researcher Gaya Herrington, a member of Earth for All's Transformational Economics Commission, decided to see if the data were still tracking today. She compared data gained in the last four decades with four scenarios from the latest version of World3.[7] One of these old scenarios assumed the world does little to change course, continuing on an economic and political business-as-usual course (BAU in figure 1.1). An updated version of the original BAU scenario assumed twice as much natural resources like fossil fuels (BAU2). A third scenario assumed massive, comprehensive technology innovation (CT) to solve some of the problems encountered when approaching global limits, like food availability. And the fourth scenario explored a route to stabilize the world (SW) by shifting priorities away from growth in material consumption and instead investing in health and education, reducing pollution, and using resources more efficiently.

Herrington's study is a reminder that *The Limits to Growth* was intended to explore several paths toward different possible long-term futures, not to make one prediction. She found that the first three scenarios most accurately tracked the actual data. This tells us two things. First, as Herrington put it, "The close alignment of empirical data with the scenarios is a testament to the accuracy of World3." And second, this close alignment between model and reality should set off alarm bells. The first two scenarios pointed toward collapse in the

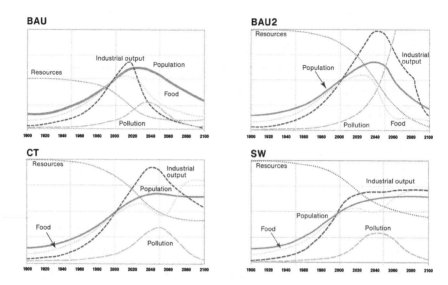

Figure 1.1. Four scenarios from *The Limits to Growth*: BAU, BAU2, CT, and SW. Graphs by Hillary Moore.

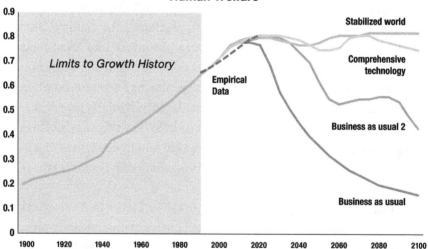

Figure 1.2. *The Limits to Growth* scenarios vs historical data from the UN Human Development Index up to 2020 plotted against human welfare variables for all four scenarios. Credit: Gaya Herrington (2021).

twenty-first century. BAU showed a world whose material consumption crashed up against planetary boundaries. When resources were doubled in BAU2, inefficient overuse continued for longer, ultimately leading to the biggest collapse due to excessive pollution. The scenario relying on technological innovation led to serious declines in resources and industrial output but not collapse. Only the fourth scenario—a large-scale transformation of societies—led to widespread increases in human welfare and population stabilization.

Love it or hate it, *The Limits to Growth* report sparked international debate about civilization, capitalism, fair resource use, and our collective future that continued many years after publication. Ronald Reagan famously attempted to discredit the report by stating: "There are no great limits to growth because there are no limits of human intelligence, imagination, and wonder."

Reagan may well be correct about limitless human imagination, but that does not change the fact that we live on a physically finite, crowded planet undergoing profound changes. It is time to start using that intelligence, imagination, and wonder to reimagine and build equitable societies where citizens can thrive and have freedom to follow their dreams within the planetary boundaries of our one and only Earth.

From *The Limits to Growth* to Planetary Boundaries

Since the publication of *The Limits to Growth* in 1972, one scientific conclusion has eclipsed all other scientific insights in the last fifty years. Earth has entered a new geological epoch: the Anthropocene.[8] This paradigm shift in our understanding of both civilization and the Earth system is as profound as Copernicus's conclusion that Earth orbits the sun or Darwin's theory of natural selection.

Geologists split deep time into units: the Jurassic, the Cretaceous, the Carboniferous, and so on. These mark out the major shifts in our planet's evolution. In 2000, Paul Crutzen, a Nobel laureate on the committee of the International Geosphere-Biosphere Programme, proposed that Earth has entered a new geological epoch, the Anthropocene.[9] This idea quickly gained momentum among the research community. By recognizing the Anthropocene, scientists

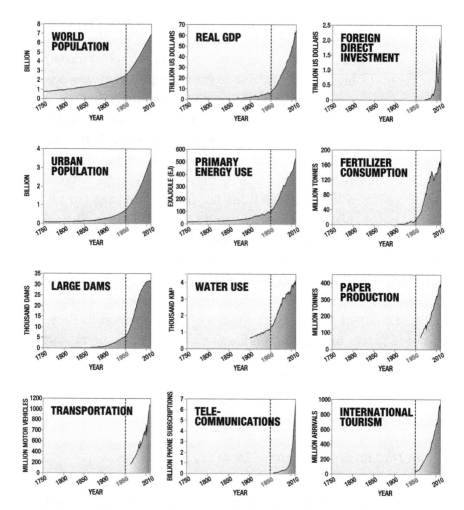

Figure 1.3. The Industrial Revolution (*left-hand panel of graphs*) began in 1750. But the destabilizing impact of the revolution on the Earth system (*right-hand panel of graphs*) only becomes apparent beyond 1950. This pattern has become known as the Great Acceleration. This delineates the Holocene from the Anthropocene. Source: Steffen et al. (2015).

acknowledge that the dominant driver of change within the Earth system is now a single species: Homo sapiens, us. Without doubt, what has happened to our planet in recent decades is absolutely unique in its 4.5 billion-year history.

The epoch we left behind, the Holocene, served human civilizations well. It began 11,700 years ago at the end of the last Ice Age. After

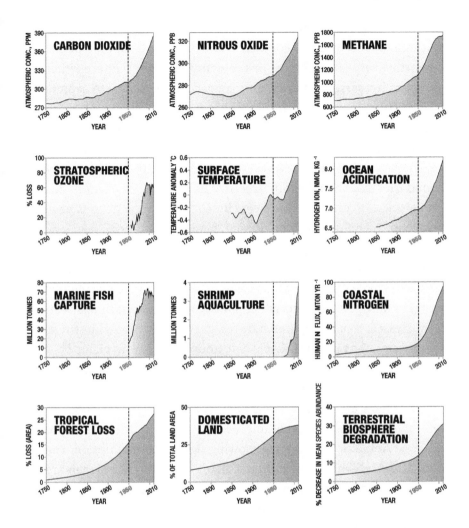

a few bumps, the climate settled into a remarkably stable rhythm. It is no coincidence that civilization emerged almost immediately. This mild climate and relative stability allowed agriculture (and the production of surplus). It has lasted 10,000 years and could have been expected to last a further 50,000 years.[10] But this is now in jeopardy. The growth of industrialized societies, largely since the 1950s, has pushed Earth out of the Holocene boundary conditions. We are in unknown territory. The explosive growth and its direct effect on Earth's life support system is best illustrated by the Great Acceleration graphs (figure 1.3).[11]

As the scientific understanding of the Anthropocene has grown, researchers have become concerned about potential tipping points on Earth—very large climate or ecological changes that are either abrupt or irreversible or both. This concern has led some to explore the conditions that keep Earth in a system state similar to the Holocene. This is worth some emphasis. The Holocene is the only state we know that will support a large civilization. In 2009, a team of researchers published a new framework identifying nine planetary boundaries that should not be exceeded if Earth is to stay within this stable state. In 2015, scientists concluded that human actions have breached the boundaries for climate, biodiversity, forests, and biogeochemical cycles (largely our use of fertilizers using nitrogen and phosphorus). In 2022, scientists announced a fifth boundary has been crossed: chemical pollution including plastics (see figure 1.4).[12] As we write, in May 2022, experts are exploring whether a sixth has been transgressed, too: a proposed new category of freshwater known as green water, the moisture held in soil around the roots of plants.[13]

The tipping point risk is now acute. Indeed, in 2019, scientists announced that an alarming number of tipping "elements"—places we know there are grave risks—are now getting "active." The Amazon rainforest is losing carbon at unprecedented rates. Critical parts of the West Antarctic Ice Sheet are showing signs of destabilization. The permafrost in Siberia and northern Canada is thawing. Coral reefs are dying. And Arctic sea ice in summer is on a downward spiral.[14] Tipping points are not some future risk later this century. We cannot rule out that Earth has recently already crossed several tipping points.

For these reasons, we can categorically define our current situation as a planetary emergency. To use a Titanic comparison, if sixty seconds remain before colliding with an iceberg and it takes sixty seconds or more to turn enough to avoid collision, this can obviously be described as an emergency situation. It is touch-and-go whether there is enough time to act. The alarms should be ringing. For our own civilization here on Earth it will at best take one generation to sail away from dangerous tipping points into a safe zone. The extraordinary turnaround must start now.

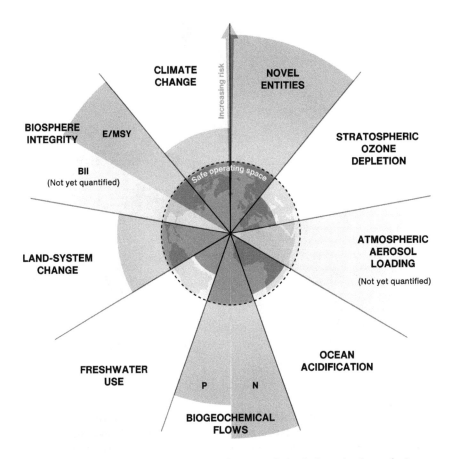

Figure 1.4. The planetary boundaries framework depicting nine boundaries that determine the state of the planet. The central area represents a "safe operating space for humanity," which provides a good chance of ensuring that Earth remains within Holocene-like conditions. Outside this area, there are profound uncertainties about how the Earth system operates. The risk of crossing abrupt or irreversible tipping points, for example, rises as Earth goes further beyond boundaries. In 2015, the planetary boundaries research team assessed that four boundaries had been transgressed. In 2022, another research team assessed the novel entities boundary, which includes plastic and other chemical pollutants, for the first time. The team proposed that this boundary, too, has been transgressed. Credit: Azote, Stockholm Resilience Centre.

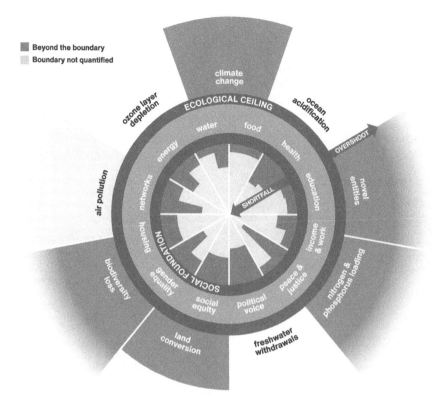

Figure 1.5. The "doughnut" of social and planetary boundaries. We have crossed five planetary boundaries, and many around the world live outside of the social boundaries. The goal is to bring humanity back into its safe operating space, represented here by the area between the ecological ceiling and the social foundation. Source: adapted from Raworth (2017).

The planetary boundaries framework has helped catalyze new thinking about risk and inspired many research groups to explore the implications for policy and economic growth. UK-based economist and Earth for All Transformational Economics Commissioner Kate Raworth took the framework and added twelve *social boundaries*—minimum standards for essentials like access to water, food, healthcare, housing, energy, and education. The doughnut-shaped graphic in figure 1.5 depicts both the planetary and social boundaries and defines a safe operating space for the human economy—"living in the doughnut," the area in which human activity does not overshoot Earth's ecological ceiling and humanity does not fall beneath the social foundation.[15] Within this space, economies focused on well-being can flourish. Far too many people around the world live below the social threshold, risking the social tipping points we will explore in chapters 3 and 4.

The Earth for All Initiative

While The Limits to Growth, the planetary boundaries and the doughnut are the scientific starting points for Earth for All, the closest things the world has to an agreed vision for civilization are the seventeen Sustainable Development Goals (SDGs) announced by the United Nations in 2015. All countries have agreed to try to meet these goals, ranging from ending poverty to energy for all, by 2030.

But some pretty big questions remain unanswered though. Are the SDGs even achievable? What will it take to achieve them? And looking beyond 2030, what are the pathways likely to deliver long-term prosperity for all on a stable planet?

The Earth for All initiative was created to build a network of scientists, economists, and thought leaders to look at these questions and explore the most plausible pathways to meet the SDGs, and beyond toward a safe operating space for humanity, toward wellbeing economies, toward life in the doughnut. The analysis provides, we hope, some useful guidance on what the priorities are, how much we need to invest, and what fundamental changes our societies and economies must embrace to increase the likelihood of success this century. In that sense, we hope *Earth for All*, the book, is a twenty-first-century survival guide for a civilization on a finite planet.

However, we must acknowledge we certainly don't have all the answers, nor can anyone predict the future. Other networks and research groups are fortunately also working on these same challenges. Looking across at their work, we can see there is broad convergence on the required urgent transformations. This gives us confidence that we are on the right track. Still, we will highlight where there is (and will always be) disagreement among experts and continued tensions that make our path forward difficult.

We developed the Earth4All model to help explore and illustrate scenarios. You will read more about the two most important scenarios in chapter 2. Each describes a plausible future. Too Little Too Late assumes societies make decisions and respond to future challenges the same way they have in the past—through incremental policy improvements. Giant Leap assumes societies recognize the shared interlinked crises and immediately start changing course through extraordinary actions in five key areas.

Our foresight analysis for this century strongly indicates that the Giant Leap's five extraordinary turnarounds can be achieved by implementing key policy goals.

- **Poverty.** Low-income countries should adopt new rapid economic growth models that secure wellbeing for the most vulnerable. A starting point is reform of the international financial system to de-risk and revolutionize investment in low-income countries. *Key policy goals: GDP growth rate of at least 5% per year for low-income countries until GDP is greater than $15,000 per person per year; the introduction of new indicators for wellbeing.*[16]

- **Inequality.** Shocking levels of income inequality must be addressed. This can be achieved through progressive taxation and wealth taxes, empowering workers, and dividends from a Citizens Fund. *Key policy goal: The wealthiest 10% take less than 40% of national incomes.*

- **Empowerment of women.** Transforming gender power imbalances requires empowering women and investing in education and health for all. *Key policy goal: Gender equity that will contribute to stabilization of global population below nine billion by 2050.*

- **Food.** To transform agriculture, diets, food access, and food waste; by 2050 the food system must become regenerative (storing vast volumes of carbon in soils, roots, and trunks) and nature positive. Local food production should be incentivized, and excess inputs of fertilizers and other chemicals significantly reduced. *Key policy goals: Healthy diets for all while protecting soils and ecosystems and not expanding the amount of land, overall, devoted to agriculture; dramatically reducing food waste.*

- **Energy.** We must transform energy systems to increase efficiency, accelerate the rollout of wind and solar electricity, halve emis-

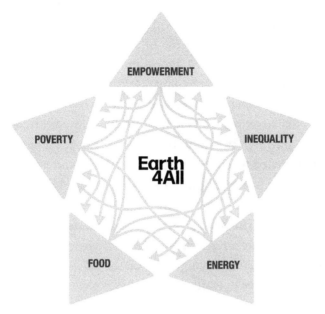

Figure 1.6. The five turnarounds are interlinked so that together they create a whole system transformation.

sions of greenhouse gases every decade, and provide clean energy to those without. This will also deliver energy security. *Key policy goal: Halve emissions approximately every decade to reach net-zero emissions by 2050.*

We called these five main solutions "extraordinary turnarounds" because they break with the trends of the past in significant ways and hold the potential to bring about real systems change. In a way, these turnarounds might form the basis of a new social contract for functioning democracies in the Anthropocene.

Chapters 3 to 7 describe in detail what these extraordinary turnarounds involve and how they can be accomplished. As you will see, they are deeply, systemically interlinked: energy influences food, and both food and energy impact the larger economic system. Removing poverty entails redistribution of wealth, which creates trust and accelerates wellbeing. And empowering women creates economic opportunity, reduces family size and inequality, and promotes healthier relationships in all societies. As Dr. Mamphela Ramphele, Club of Rome co-president and Earth for All Transformational Economics Commissioner, reminds us, "The essence of being human is to be interconnected and interdependent."[17]

For sure, implementing these extraordinary turnarounds is a daunting challenge in a world of deep complexity. But, from termite colonies to starlings sweeping through the sky, from weather prediction to the global economy, we know that seemingly unfathomable complexity can arise from a small number of simple rules or relationships.

We list three of the most powerful socioeconomic levers for each turnaround in figure 1.7. At the bottom of the pyramids are what we consider the basic policy changes within the current economic paradigm, but then move upward to the bolder policies that really define a new economic paradigm fit for the Anthropocene. At the top of the pyramids are the levers that really deliver the transformation to a new

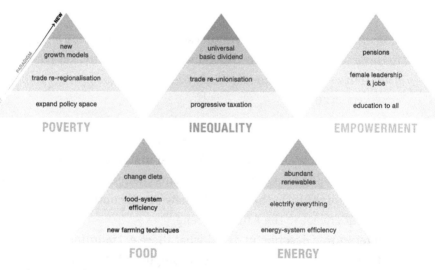

Figure 1.7. The Earth4All paradigm shift illustrated as five triangles. Each turnaround has key levers that will drive a disproportionate impact. Reading from the bottom of each, we begin with the economic solutions within the current paradigm. At the top are the proposals that really are a Giant Leap. They move us into a new paradigm.

economic paradigm, which some have called "wellbeing economics." Here you find some of the big ideas that lock in Earth for All: It is only when we pull these bold levers in the Earth4All model—early and strongly—that we see accelerated transformation toward a sufficiently fair, just, and safe world by the middle of this century.

You—yes, you—can explore other scenarios and solutions with the Earth4All model, and we encourage you to do so. It is available as a simple online tool. See appendix 1 for instructions.

You may have noticed that a number of issues are not explicitly the subject of their own extraordinary turnaround. Where is governance? Surely this needs an overhaul. Where is health? Or exponential technologies like automation and artificial intelligence? What happens in these domains will of course alter humanity's future on Earth. These issues are common threads woven into our scenarios. If you want detail, read more about these issues in the accompanying Deep Dive papers from the Transformational Economics Commission.

You may also have noticed that we haven't given material consumption its own turnaround. Instead, this, too, is a common thread through all the turnarounds, as it must be, because the scale is eye watering. Since 1970, natural resource extraction has tripled. In 2020, Earth crossed a grim threshold: the weight of concrete, steel, plastic, and all other materials produced by humans surpassed the weight of all living biomass on the planet.[18] Today, with almost eight billion people on Earth, we produce 530 kilograms of cement and 240 kilograms of steel per person per year.[19] After clean water, concrete is the second most consumed product on Earth. It is no surprise then that production of steel, iron, and cement accounts for some 14% of carbon dioxide emissions globally.

Demand is growing. But it does not have to. Whatever future we build, it will require materials. If we value our future on a stable planet, then we simply have to do more with less. Ultimately, governments need to incentivize a rapid transition to circular economies. Small changes to building codes could shrink demand for steel and cement by around 25%. Steel and aluminum are already two of the most recycled materials on the planet, but where we must use new materials, we can change the production system. Using hydrogen, instead of coal, to produce steel, for example, cuts emissions an incredible 97%.[20]

But there are important issues around fairness when it comes to consumption. Consumption is not evenly spread around the world. The twenty wealthiest countries use over 70% of these resources. And the fastest-rising source of greenhouse gas emissions worldwide—by far—are the richest 1% of societies. Overconsumption is a *systemic* challenge: economies optimize consumption at the expense of social cohesion and human and planetary health. Here we tackle consumption head-on. Each turnaround aims to reduce unfair and unnecessary material footprints. Some of this is tackled through progressive taxation, in other places, a Citizens Fund can help reduce unsustainable consumption, and redistribute wealth more fairly. By essentially reducing the material consumption of the wealthiest in societies, and by adopting smarter ways to provide what people really need, we can make more room for Most of the World to have their fair share of resources.

Consumption and a country's gross domestic product (GDP) are linked. Over the last couple of generations, after World War II, GDP has evolved into becoming the preferred way to determine the health of an economy. This is despite the fact that it does not measure health or wellbeing. GDP is simply a measure of the total activity level in the economy, measured in dollars per year. It is nothing but the total output of goods and services produced in a year, multiplied by the price per unit of output. In poor economies, with low labor productivity, rising output initially leads to higher wellbeing. But above a certain income, this no longer holds true. Many studies show wellbeing plateaus as GDP grows. While, yes, people can buy things, they have to contend with clogged arteries from poor diets, clogged cities with SUV-infested streets, and clogged lungs from air pollution. At this stage, rational government policies would shift focus away from growth and rather seek increased wellbeing for the majority.

Generally, political leaders should be agnostic about growth. It really depends on what is growing. Low-income countries need to grow their economies—especially since this can be done sustainably. And if we succeed in solving the energy and food challenges, this will lead to GDP growth—and this time, this type of growth will lead to higher wellbeing in the long run. So instead of a myopic focus on the last month's figure for GDP growth, political leaders and their voters should ask: Is the economy optimized to improve the lives of the majority? Is the system perceived as reasonably fair? Is the economic growth *responsible growth*? Few countries could answer yes to these questions. And people know it.

People Support Economic Systems Change

What we are proposing will require unprecedented economic shifts in a single generation—actually, within a single decade. Are citizens ready to change? Beyond the protesters on the street catching media headlines, is there a widespread mood for broader systems change in societies? Are citizens aware of the scale of the risks we face in future? And do people want to act? Are they ready for a new economic system that truly values wellbeing for everyone? That is, a truly equitable future?

We conducted a global survey (G20 countries[21]) to find out (see chapter 9). The results show overwhelming public support for policy-makers to deliver systemic economic changes necessary for building a nature-positive, zero emissions, and equitable future for all. The findings should provide leaders with the bottom-up public backing to implement policies in line with Earth for All goals much faster.

Momentum is building for change. As we slide deeper into the twenty-first century, people everywhere have been impacted by frequent economic crises, pandemics, wars, floods, fires, and heat waves. But too many people see no viable way to achieve economic security. Even in the richest societies the world has ever known, many feel economically insecure, left behind, or constantly worried about being left behind. And in the poorest countries, they watch rich nations pulling up the drawbridge around the fortress: "No entry." The 2008 global financial crisis so blatantly showed that banking profits are in private hands, but the public is expected to pick up the cost of the losses. The conventional growth model seems as bankrupt as the conventional economics of efficiency and austerity. No coherent solutions are currently in sight.

We wrote *Earth for All* to provide a fresh, credible, consistent story of how to transform the global socioeconomic system during the next fifty years based on scientific knowledge and illustrated using quantitative system dynamic modeling. The results were reviewed by the multidisciplinary experts from across the globe in the Transformational Economics Commission, and weaknesses identified and discussed. We are not presenting an exhaustive list of solutions. Rather, these are some of the ideas that, in our opinion, could have the most leverage in the shortest time. We hope they spark debate. And we invite better ideas!

In *Earth for All* we present an aspirational, stubbornly optimistic guide to the future. But how likely are we to get there? That, dear reader, depends on what you do next.

2

Exploring Two Scenarios
Too Little Too Late or Giant Leap?

Scenarios are stories about the future that can help us make better decisions today. Each one describes a plausible development path over the coming decades or longer. But they are not forecasts. They don't predict the most likely future as, say, a weather forecast would predict the most likely weather. Rather, scenarios give answers to important "What if...?" questions:

"What if the world continues on with high levels of inequality and planet-busting overuse of materials?" Or, "What if governments change course?" or "What if new renewable technologies become much cheaper quicker?"

Scenarios are useful when one is faced with large uncertainties about the future. But to be really useful, each scenario, however, has to be internally consistent, and derived from a set of valid assumptions drawn from data and observation over time. We know that higher levels of education in a country, for example, will lead to higher incomes and smaller family sizes, so a useful scenario on social changes should reflect that. Often researchers develop a group of scenarios to reflect possible futures and to evaluate their relative attractiveness. In this way, scenarios can help people not only to plan for an uncertain future but to actively create and shape it.

We use the Earth4All model to generate internally consistent scenarios for global development toward the year 2100. The assumptions in the model are drawn from real-world data from 1980 to 2020, and the model is designed to reproduce the observed historical development patterns of population growth, education, economic growth, greenhouse gas emissions, and all its other variables across all ten regions during that time. And even though Earth4All is a grossly simplified description of the real world, it does indeed track major global and regional trends over the last four decades reasonably well.[1] This gives us some confidence in its utility as a tool to describe possible future scenarios in an internally consistent manner. We start the model in 1980, let it recreate historical trends to 2020, and then run it into the future under different policy assumptions, to explore how the ten regions could evolve over the century to 2100, depending on what decisions people collectively make.

The role of the Earth4All model is to show consistent pictures of possible futures—to help evaluate the potential consequences of various alternative decisions, and to discover which system changes are likely to have a big effect and which are likely to make little impact. The model also gives us an idea of how much these changes will cost and what level of investments are needed to reach a certain level of wellbeing by a certain time.

We explored many scenarios but present just two in this book. Too Little Too Late reflects our current trajectory, where societies keep boasting and bumbling about "sustainability" while muddling through. In Too Little Too Late, most countries make piecemeal, incremental progress toward ending poverty and stabilizing the climate, but do not really deal with the elephant in the room: inequality. Will "muddling through" produce safe passage to the end of this century, or will deep fault lines emerge within and across societies, throwing democracies into disarray and risking deep destabilization? In Too Little Too Late, the latter possibility appears more likely. In this scenario, social trust declines as the richest 10% and bottom 50% continue diverging, while societies and nations turn against each other, competing for resources. There is too little collective action to limit the immense pressure on nature. Earth's life-supporting systems like forests, rivers, soil, and climate keep deteriorating, and some systems move closer or cross irreversible tipping points. For those in poverty, Indigenous peoples, and wildlife, this is a steady "stairway to hell."

As our second scenario, we chose Giant Leap, which illustrates the effects of the powerful and immediate implementation of the five extraordinary policy turnarounds. Its passage through the century is driven not by tinkering at the fringes but by fundamentally reconfiguring economies, energy systems, and food systems. It's a major upgrade. A reset. An essential reboot of civilization's guiding rules before the system crashes. Due to inertia in economies and in the climate, the main impacts of any action taken today are often not seen for years in economies and decades or centuries in the climate. We believe that nothing less than a "giant leap" now is needed if humanity is serious about turning around from the current trajectory and getting on a new track to a sustainable world by 2050. The Giant Leap scenario spells out the details of a new type of economy fit for the Anthropocene—an economy that removes poverty, promotes social and environmental wellbeing, and measures its progress by how well people and the planet thrive.

The Giant Leap scenario requires active governments willing to reshape markets and drive long-term visions for societies. It is our fundamental belief that neither individuals nor markets can do it alone. Turning our economy around is a collective action problem. But how to create the conditions for governments to become more active, particularly in democracies? A very basic condition is trust. Two important features of the Earth4All model are the Average Wellbeing Index and the Social Tension Index. The first provides an indication of people's quality of life over time. (See the What Is Wellbeing? sidebar.) The Social Tension Index indicates the governability of a region and rises when there is decline of wellbeing, trust, and equality within societies.

What Is Wellbeing?

The Wellbeing Economy and Earth4All's Average Wellbeing Index
A growing number of economists, policymakers, business people, and other changemakers have been developing new frameworks for organizing economies and measuring societal progress. This new economic thinking has brought us concepts

like the caring economy, the sharing economy, and the circular economy. It has incorporated ecological economics, feminist economics, doughnut economics, and other new ways of looking at what creates and sustains prosperity while also protecting the planet. These are not just competing buzzwords for the same concept; rather they emphasize different aspects of alternatives to our current linear, neoliberal, growth-at-all-costs economic approach.

The transformed economy envisioned by the Earth for All project adopts elements of all these frameworks and aligns with the comprehensive framework known as a wellbeing economy. The Wellbeing Economic Alliance (WeAll) describes this framework as "one that serves people and the planet, rather than people and planet serving the economy. It does more than 'move money around' but delivers good lives to people."[2] WeAll describes the core needs for human wellbeing as:

- **Dignity:** Everyone has enough to live in comfort, health, safety, and happiness.
- **Nature:** A restored and safe natural world for all life.
- **Connection:** A sense of belonging and institutions that serve the common good.
- **Fairness:** Justice in all its dimensions is at the heart of economic systems, and the gap between the richest and poorest is greatly reduced.
- **Participation:** Citizens are actively engaged in their communities and locally rooted economies.

Setting wellbeing as the ultimate goal for economies means meeting human needs and capabilities while acknowledging the biophysical reality of a finite planet. This is precisely the goal of the Earth for All project, and it is reflected in the Earth4All model's chief measure of progress—its Average Wellbeing Index, which simulates wellbeing annually.

The index provides an alternative to GDP—an economic progress indicator that has widely and wrongly been used as a proxy for human wellbeing. Researchers have found that beyond a certain threshold of GDP per person, further rises in GDP are not associated with further increases in life satisfaction. A recent study from the Earth for All initiative confirms that the fulfillment of human needs and aspirations does not increase considerably when GDP per person grows beyond a threshold of some $15,000 per person per year).[3] One reason is that growth in GDP normally leads to higher negative environmental side effects.

Despite its dominance as an indicator, GDP was never intended to measure a society's overall health, only to measure its activity level. Over the last several decades, the dire consequences of treating GDP growth as a singular goal have shifted attention toward economic measures that emphasize wellbeing. These frameworks are inherently pluralistic, and take local circumstances, value systems, and traditions into consideration. They are united in the acknowledgment that human wellbeing requires a broader conception than merely maximizing income and consumption.

The goal is not to eliminate GDP as an accounting metric but to move beyond GDP to use wellbeing as the central guide to societal progress. A viable indicator must include the interdependence of human wellbeing and a healthy planet. Human needs are universal, but how they are satisfied depends on cultural circumstances.[4]

Hence, the Earth4All Wellbeing Index builds on the WeAll principles, and quantifies wellbeing through these select variables in the model:

- **Dignity:** worker disposable income (after tax)
- **Nature:** climate change (global surface average temperature, in Celsius)
- **Connection:** government services indicated by spending per person, i.e., to institutions that serve common good
- **Fairness:** the ratio of owner income after tax to worker income after tax

- **Participation:** people's observed progress (previously improving wellbeing) and labor participation

The Average Wellbeing Index is calculated annually, based on the variables listed above, for each of the ten regions in the Earth4All model. It reflects the wellbeing of an average person. When the Average Wellbeing Index declines, people suffer and get angry, leading over time to increasing social tension and the rising risk of political instability and even revolution.

The links between gross inequality and unstable societies are well-known, as pointed out by Earth for All Transformational Economics Commissioners Richard Wilkinson and Kate Pickett.[5] In societies with large and growing economic inequality, unless checked, the wealthiest have a disproportionate influence over governing institutions. This undermines trust in the governance system and opens the doors for corruption. Inequality also leads to lower wellbeing in societies: It reduces social cohesion and heightens status competition.[6] Over time this drives the Social Tension Index upward. When the index has been rising over a longer period, then we can expect deep rifts within societies and an unhealthy us-versus-them dynamic emerging that can be easily exploited by politicians. If the Social Tension Index rises too far, societal collapses cannot be ruled out. We use this term to mean societies enter a vicious cycle where rising social tensions lead to a decline in trust causing political destabilization. This in turn leads to stagnating economies and wellbeing falls further. Governments struggle to regain trust, making it increasingly challenging to make coherent long-term decisions.

Both our scenarios are at the level of macroeconomics, looking at economies as a whole. But what do these macro-system changes actually mean to real peoples' lives at the micro everyday level? To make these two scenarios tangible, we developed four characters—all girls, born on the same day in early August 2020—and imagined their trajectory through each one. Shu was born in the Chinese city of

Changsha; Samiha in Dhaka, Bangladesh; Ayotola in Lagos, Nigeria; and Carla in the United States. These are not real people but more like avatars that highlight what it is like to live in the Too Little Too Late and Giant Leap worlds. We chose four girls in order to better compare across regions, scenarios, and opportunities.

Like 1.4 billion other people on Earth, Samiha and Ayotola were born into vulnerable informal settlements in their cities. And like 3 to 4 billion people on Earth, their families exist on less than $4 per day. Shu and Carla's families are better-off economically. Shu's mother is a teacher and her father an accountant in Changsha. Carla's parents moved to California from Colombia for the economic opportunities in the United States. Her mother stays at home to look after the three children, and her father works in the restaurant industry. We will follow their journeys from 2020.

A Brief Review of 1980 to 2020

Both scenarios build on the major trends since 1980. In this period, the largest economies rapidly adopted neoliberal policies of privatization, deregulation of markets, globalization, free trade, and public commitment to reduced government spending. In the high-income countries, corporations gained power and trade unions' negotiating strength dwindled. Declines in public spending weakened economic security. The gap between rich and poor within countries grew ever wider, and rising inequality undermined public trust in political institutions in many areas.

The global population continued to rise, and while the fraction of poor declined, absolute poverty lasted well into the century. The world economy (measured by its GDP) continued to grow, albeit more slowly than in the twentieth century. The financial sector (the banks, money markets, hedge funds, and private equity firms, among others) expanded like a balloon, growing in size and importance to the point where the sector became a major driving force of many countries' economies. In 2008, the balloon burst spectacularly, destabilizing economies and societies. Subsequent reforms appear to have reduced the fragility of the finance sector but not its dominance over our economy or the focus on short-term profits over long-term value creation and human progress.

Greenhouse gas emissions rose rapidly, and global average temperature exceeded 1°C above preindustrial times by 2015. This was a landmark for the planet. The Holocene epoch was defined by its stable temperature, never rising or falling more than 0.5°C in 10,000 years. This was a landmark, too, for civilization, which had thrived within the Holocene boundary conditions.

Economic inequality within countries kept growing throughout the four decades. New digital technologies disrupted traditional industries and their workforces. As part of the process of globalization, companies gravitated to cheap labor and lax regulations, abandoning many workers in the global North. The Social Tension Index within most regions rose steadily, with implications for effective governance.

Investment in clean energy technologies such as wind and solar power and electric vehicles improved gradually until these technologies finally reached cost competitiveness (after subsidies) around 2015. Now, their costs and performance compete favorably with fossil fuel alternatives as we enter deeper into the 2020s.

Carla's parents were born when Reagan was president, and arrived in a country where unions were seen as lazy and corrupt and selfishly out to destroy the United States' competitiveness. In fact, globalization and technological innovation were significant drivers of the shrinking US manufacturing industry. Shu's parents were born in China in the period when the communist government started under Deng to implement market reforms and opened up to trade and investment, while maintaining active control with the direction of economic growth through five-year plans and a competent central government. This led to dramatic growth over four decades, pulling hundreds of millions of people out of extreme poverty.

In Bangladesh, where Samiha's parents were born, and in African countries such as Nigeria where Ayotola's father and mother grew up, economic growth was slower. These countries often relied on international financial institutions but became burdened by neocolonial arrangements and debt. This often resulted in a lack of investment in building domestic manufacturing jobs and sectors.

Before we jump into the future, we want to emphasize that societies, ecosystems, and economic systems are dynamic. The forces that drive them interact with one another and this can lead to surprises. Likewise, the Earth4All model is dynamic: Different variables like population, investment in public capacity, economic output, energy demand, or food production interact with each other. If one variable grows, the impacts ripple through the model system eventually affecting the world economy and the planet's life support systems. This allows us to explore in a rudimentary way what may happen to the global population when economies grow and how this affects food supply and emissions. We can also introduce global and regional policies into the model that help promote equity, trust, and system resilience: What happens if governments implement a small or large wealth tax, for example? Or invest more in technological innovation?

Scenario 1: Too Little Too Late

This scenario shows the potential consequences of continuing world development along the same dynamics as 1980 to 2020 (see Figure 2.2). The overall global result is a somewhat slowing population growth and world economic growth to 2050 and beyond, but also declining labor participation rates, declining trust in government, a steady increase in the ecological footprint, and rising loss in biodiversity.

The regional upshot in the coming decades is persistent poverty in Most of the World and destabilizing inequality in the rich world. Some of the Sustainable Development Goals are met, and there is some progress toward living within planetary boundaries. But overall, there is a dramatic rise in the Social Tension Index (see figure 2.1), which slows down the deployment of new solutions. The economy is not turned around but churns on along as it has for the last few decades. Although the scenario does not result in an overt global ecological or climate collapse this century, the likelihood of *societal collapse* nevertheless rises throughout the decades to 2050. This is a

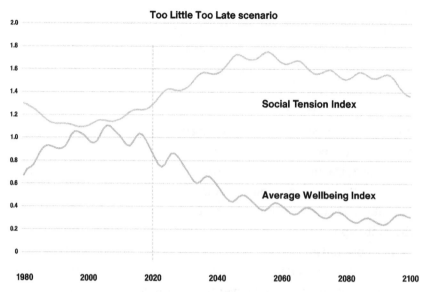

Figure 2.1. A declining Average Wellbeing Index drives the Social Tension Index to rise over time to a peak in mid-century, reflecting worsening distrust and societal fragmentation.

result of deepening social divisions both internal to and between societies and rising environmental damage. The risk is particularly acute in the most vulnerable, badly governed, and ecologically vulnerable economies.

Scenario Too Little Too Late: The Decisive Decade 2020–2030

The four girls, Shu, Samiha, Ayotola, and Carla, are born at a time of deep uncertainty. A pandemic is sweeping the world. Cooperation between countries is limited. Levels of inequality not seen since the Great Depression[7] have led to a rise in populism and authoritarianism in many regions.

By 2030, our four girls are bright ten-year-olds growing up in a turbulent world. In the outskirts of the Chinese city of Changsha, Shu's school is often closed due to air pollution across the city. When five years old, Shu suffered pneumonia and later from asthma. Her parents are saving up to move her to a "bubble school" where air is filtered. Across the Pacific, Carla's parents face a similar dilemma.

Droughts and wildfires have become more intense in the Los Angeles area. For several months of the year, the air quality is so bad it is risky to go outdoors. Carla is taught to use water sparingly so there is enough for everyone. Samiha's parents in Bangladesh are also worried about water but for the opposite reasons. The Bangladesh government is now spending so much money on flood defenses it has to make difficult decisions about investment in hospitals or education. Samihi and her siblings have to drop out of school to help bring in money. The settlement becomes more and more crowded due to increasing numbers of climate refugees arriving on a daily basis. In Africa, by the end of the decade, Ayotola's home city of Lagos is now home to 20 million people, making it among the largest cities on Earth. But economic opportunities here are still limited, and her parents dream of moving to Europe.

In Too Little Too Late, countries have agreed to set *goals* to curb climate disruptions, but policy *actions* remain far too weak to meet the Paris Agreement. The energy transition begins to accelerate nevertheless, mainly through market forces and thus limited to what is profitable given limited subsidy and carbon pricing schemes. The share of renewables in the energy mix rises slowly but steadily through the decade. India and other low-income economies begin their rapid growth phase through a mixture of old technologies—fossil fuels and new technologies like wind and solar. Renewable energy tends to be used in addition to fossil fuels, as energy demand rises briskly. The promised climate funding from high- to low-income countries never materialized. Old trade agreements still hinder sufficient health and green investments in low-income countries.

Investments in profitable technology continue and create revolutions in robotics, the Internet of Things, 3D printers, and artificial intelligence. These drive industrial disruptions in most sectors. Low- and mid-skilled labor loses out as the economy is optimized for gig work and zero-hour contracts. Conventional manufacturing struggles to survive in rich regions. While new industries emerge, they are often located in different geographies, and retraining is not prioritized. In the transition, labor is viewed as expendable, forcing more people to live economically precarious lives. Real estate prices keep rising as ever more speculative capital goes in search of easy profits rather than supporting and driving a new economy. Living in large cities becomes increasingly unaffordable for many. In all regions, within-country inequality keeps rising, as the top 10% capture most of the gains.

Climate injustice is more visible with the poorest bearing the brunt of impacts. This contributes to a rise in social problems such as polarization and tensions. Demonstrations erupt frequently, often driven by violent nationalists. In democracies, people often vote out of anger[8] and express contempt for politics by swapping parties often.

At the UN General Assembly in 2030, despite absolute poverty reaching its lowest level in history, the world has fallen short of meeting the Sustainable Development Goals. The economic differences between high- and low-income regions keep increasing. The global temperature is now 1.5°C above preindustrial levels. The world continues to witness an ominous rise in intense and deadly heatwaves and other extreme weather.

Scenario Too Little Too Late: 2030–2050

Extreme poverty falls to its lowest level in 2050, but there is still a long way to go until it disappears. Yet within-country inequality has dramatically worsened. Inequality causes rifts between the super-rich, with most access to power, and the poorest half of society. This has implications for the stability of democracies, and some countries become increasingly difficult to govern. There is private wealth but public austerity. Public spending on health and education suffers along with pension schemes. This also prolongs population growth, as people in many places view larger families as security for old age.

At the same time, the energy and technological innovations continue apace: better and cheaper solar panels, buildings, smart grids, batteries, electric vehicles, and so on. But in high-income areas, the deployment of long-term systemic solutions for sustainable energy and food systems are delayed or scaled down repeatedly. This is due to social and legal conflict and weakened governments. Whereas in low-income areas, there is lack of climate funding for such large upfront investments. Despite a lot of talk on climate and sustainability, the shift to a circular economy is sluggish and haphazard. The construction industry keeps building roads, railways, skyscrapers, ports, and airports using cement and steel from unsustainable production processes, and governments make little effort to incentivize smaller cars, smaller homes, and smaller fridges and freezers.

Eventually greenhouse gas emissions plateau in the 2030s and start falling. New renewables have for some time now been cheaper than coal, oil, and gas for electricity production. But fossil fuels persist in many industries including steel, concrete, plastics, shipping, aviation, and long-distance trucking. In most regions, the economic transformation has had a devastating impact on labor, with some industries contracting and disappearing while new industries emerge. Governments leave the labor forces in these regions to fend for themselves to a large extent. This creates deep and rising economic anxiety, as boom-and-bust cycles continue.

There is some good news. Air pollution is falling across China, India, Bangladesh, and the rest of Asia as coal-fired power stations close due to the rise of wind and solar and the switch to electric vehicles. This is positive for Shu and Samiha. But temperatures keep rising because greenhouse gases are still being emitted, albeit at lower levels. Across the world, record high temperatures are constantly being broken. More people are living within zones where outdoor temperature extremes now regularly exceed human tolerance.

Diets have shifted predominantly to the Western industrial diet that promotes obesity. This is driven by cheap, heavily processed foods from large agro-industrial companies. Food waste remains a major concern, and high grain-fed red meat consumption across the world means the agricultural sector is still a major driver of greenhouse gas emissions and biodiversity loss.

Africa is slowly progressing toward prosperity during the 2030s but picks up speed in the 2040s as ever more women enter the paid workforce. The savings rate eventually increases, and regional investors gain more funds. But terrible droughts and extreme climate events incur huge costs. And investment in education of girls has not rolled out fast enough to substantially slow population growth. The patriarchal social systems continue to undermine gender equity in allocation of resources and decision-making powers. By 2050, population in Africa south of Sahara reaches 1.6 billion, up from 1.1 billion in 2020. The income per person never reaches the acceptable threshold level of $15,000 per year.

In the decades leading up to 2050, economic precariousness and stagnating median incomes magnifies social tensions in Most of the World's regions. On top of this, climate migration is reaching critical levels and global health pandemics are on the rise. This fuels the spread of populist and autocratic leaders, with the risk that these leaders undermine stable governance and democratic values. Ongoing corruption reduces trust further. The risks of breakdown into smaller states, enmeshed in perpetual conflicts, remain high. Competition for common resources such as freshwater intensifies. Across ten regions of the world, our Social Tension Index, which estimates risk of societal collapse, climbs into the danger zone. This keeps destabilizing policy, which becomes characterized by abrupt starts and stops. It makes the transitions in inequality, food, energy, health, transparency, and law much, much slower than they otherwise could have been.

Scenario Too Little Too Late: 2050 and Beyond

Shu, Samiha, Ayotola, and Carla celebrate their thirtieth birthdays in 2050. Shu is now a hydrology engineer working on major projects to protect China's water supply, but floods are frequent in some areas and droughts in others. This threatens food security and the economic security of hundreds of millions of people. Mass migration in mid-century creates a housing, employment, and food crisis that escalates into a conflict. Carla is an office manager in a successful architecture business but decided to leave the extreme heat of southern California and moved north to Seattle. But she now feels the fires and heat are following her. Her brother has the same job and qualifications yet is paid much more than her. Because of her six-figure student loans and expensive rent, she lives paycheck to paycheck.

In Bangladesh, Samiha has three children but lost her job in the clothes factory as the city of Dhaka is progressively being abandoned to the flash floods by the people with the means to move inland. Means or not, Samiha knows that she will also soon have no choice, and will have to try to escape the increasingly frequent floods and heatwaves. She often wonders where she will be in a year. Ayotola left school at fourteen and married the son of a family friend. They have four children, but can only afford to send the boy to school. Ayotola does some sewing at home to make money for fish, meat, or beans to go with their ugali. The women have no experience of living on a planet without climate extremes.

The Social Tension Index has been rising since 2020 across most regions. It is not a steady rise, rather discontent comes in waves linked to how inequality and economic cycles affect wellbeing (see figure 2.1). Societies become less able to put in place long-term resilient solutions to deal with immediate crises. However, significant shocks keep amassing from many directions. There are migrations as countries near the equator become increasingly too hot to live in. Trade wars erupt as regions fight for ownership of knowledge, market shares, and resources. Supply chains falter due to extreme events. Government spending is increasingly spent on crises and adaptation, leaving less for long-term social and economic development. Soil quality is falling, affecting yields and creating food price volatility.

Global population peaks at around 9 billion around 2050 and then starts to decline toward the end of the century. As average GDP per person goes above $10,000, women everywhere choose to have fewer kids (see figure 5.3). The high-income countries provide ever-increasing material standards of living and consumption to the majority of their citizens. This is due to better technologies and an ever-higher productive capacity per person. But wellbeing is falling, due to falling social cohesion, status competition, and weak collective action to solve the biggest challenges facing societies.

In this scenario, the world misses climate targets set out in the Paris Agreement. Earth crashes through the 2°C boundary around 2050 and reaches a catastrophic 2.5°C before 2100.[9] It is likely the

Earth system will have passed several critical thresholds as a result of the escalating temperature, but, while terrified scientists keep reporting on the acceleration of meltdown of the West Antarctic and Greenland ice sheets, there is no one big bang when Earth falls off a cliff. The losses of the Amazon rainforest get worse every year, as more dries out and becomes savannah. Wildlife is lost, and extinctions of insects and birds accelerate. People have gotten richer, but the natural world is getting greyer year by year through a series of local breakdowns. Civilization has lost its greatest foundation: a stable and resilient Earth system.

A steady stream of extreme events has become the new normal. Only a few of the oldest people on Earth remember a stable climate. Government investment in adaptation now forms a significant proportion of government spending. On paper this contributes to GDP growth—but countries are "running to stand still." Some are experimenting with geoengineering to protect their populations. The climate scenario is deeply alarming, and a major contributor to societal and economic turbulence. However, the climate emergency is still unlikely to be the primary cause of *societal breakdowns*: the culprit here is likely to be both within-country and between-country inequality. Several societies start to unravel after 2050, fracturing into smaller states, often due to conflicts accelerated by climate change, as humanity enters ever deeper into the second half of the twenty-first century.

Figure 2.2. The Too Little Too Late (TLTL) scenario quantified as 4 time graphs visualizing the global developments in this scenario from 1980 to 2100. These curves are all relative to 1980 values and highlight the dynamics between them. Global population grows from 4.4 billion in 1980 and peaks at 8.8 billion in the 2050s before declining slowly. Income per person keeps rising from 6,000 $/year through to 42,000 $/year in 2100 as shown on chart 2. Chart 3 shows that carbon dioxide emissions and crop use per person remain high throughout the century, which drives global warming to around 2.5°C by 2100. Earth moves further beyond critical planetary boundaries. Chart 4 shows different components of wellbeing: the global Average Wellbeing Index declines through most of the century (mainly due to increasing inequality and worsening conditions in nature). "$" means US dollars (USD) at constant 2017 prices using purchasing power parity (PPP) rates. Source: E4A-global-220501. The model and data are downloadable from earth4.life.

1. Main trends
Too Little Too Late scenario

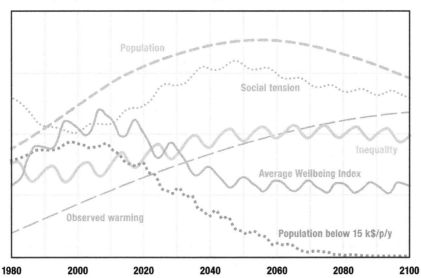

2. Human footprint
Too Little Too Late scenario

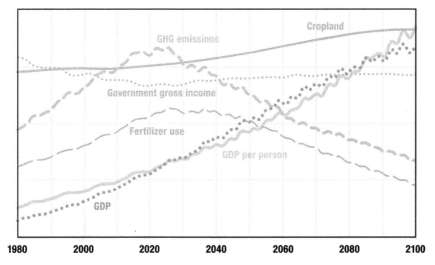

3. Consumption
Too Little Too Late scenario

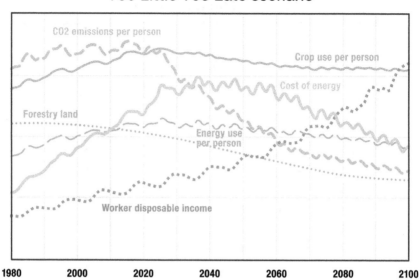

CO2 emissions per person

Crop use per person

Cost of energy

Forestry land

Energy use per person

Worker disposable income

1980   2000   2020   2040   2060   2080   2100

4. Wellbeing
Too Little Too Late scenario

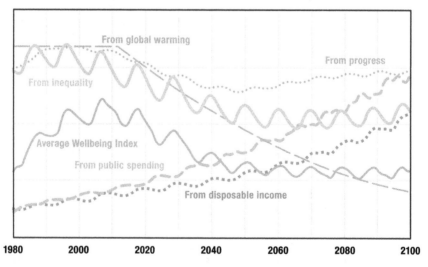

From global warming

From progress

From inequality

Average Wellbeing Index

From public spending

From disposable income

1980   2000   2020   2040   2060   2080   2100

In 2080, Carla still works long hours and knows she won't be able to afford to retire for another decade. Her sedentary lifestyle and processed food diet, as well as the frequent heat dome events over the US West Coast, make her vulnerable to health conditions. She dies of diabetes at age 65. Samiha lives in a settlement camp just outside of Dinajpur. She lost two of her three children to pandemics that swept the settlements in the last decade, and her husband died in a raid. There is no employment and limited food and safe drinking water.

Shu is by now a legend among hydrologists due to her flood-management skills. She occasionally still teaches Chinese students as professor emeritus, as she knows how badly these skills will be needed in her region in the years to come. She worries about the decline in public authority in her country. In Lagos, with deadly floods becoming more frequent every decade, Ayotola and her husband abandoned their house to the rising waters, like many others living in poverty. The population displacement contributed to rising tension across the country. Multiple governments over the past ten years have failed to appease voters. This, in turn, fostered more extremism, religious violence, and populist governments ruling by fear. Extreme weather, heat, and tropical storms across Africa are common and damage crops leading to food price volatility. Those who have the means moved to Europe and the Americas in search of better living conditions, though even in the wealthiest nations, economic insecurity and environmental shocks are the new normal.

Scenario 2: The Giant Leap

Scenario Giant Leap: The Decisive Decade 2020-2030

The four girls, Shu, Samiha, Ayotola, and Carla, are born on the same day in 2020, at the dawn of unprecedented transformation. Early in the 2020s, nations agree to start transforming international financial institutions such as the World Bank, International Monetary Fund, and World Trade Organization. Their mandates are dramatically shifted to support green transition investments in climate, sustainability, and wellbeing, rather than just economic growth and financial stability in a narrow sense. By the end of the decade, these changes have greatly expanded financial resources for low-income countries and improved access to investments in renewables and green industries. But most important, these changes have made it possible for competent and active governments to raise the wellbeing of their people, through investment in education, health, and infrastructure. The exponential technologies of solar, wind, batteries, and electric vehicles are drastically driving down the share of fossil fuels in the energy system. New development and trade models replace the dysfunctional current system that previously perpetuated historic inequities between countries (see chapter 3).

Economic inequality becomes widely acknowledged as deeply polarizing and a threat to political stability and human progress. Following shifts to wellbeing economies in Finland, Iceland, and New Zealand, other nations follow suit. There is a broad shift in attitudes in all regions to support the principle that the richest 10% should take less than 40% of national incomes. This is based on the recognition that—whether wealthy or not—fairer societies function better than unfair societies. Different regions respond with a different mix of policies. Progressive income tax ensures the wealthiest contribute more. Wealth taxes introduced in all regions, along with the closing of tax havens, address runaway wealth inequality. An international corporation tax (agreed in 2021) provides additional income for redistribution and investment by active governments seeking common prosperity, and is adjusted and harmonized every five years. And public investment in science and research is rewarded, for example, through acknowledgment of intellectual property and stock co-ownership to the public.

These new revenues allow governments to expand unemployment benefits (essential in a time of economic transformation) and pension schemes for all, particularly women. Gender equity improves along with a sharp increase in investment in education, jobs retraining, and health.

More countries adopt a Universal Basic Income (or similar) to help fight inequality, particularly as a stimulus during major shocks (like another pandemic), which then turns into regular transfers. This gives people economic freedom to retrain as some industries contract while others grow. As citizens of Earth, more feel they have a fair share of the wealth generated from the world's commons: Earth for All!

This principle evolves into Citizens Funds that pay a universal basic dividend (UBD), where industries pay for the use of common resources (for instance, from land use or ownership, financial assets, intellectual property rights [IPR], fossil fuels, rights to pollute, resource extraction of other materials that can be considered a resource commonly owned by all in society) into a Citizens Fund. This revenue is then distributed back to all citizens in a country equally. The dividend phrase, writes Earth for All Transformational Economics Commissioner Ken Webster, reflects that citizens increasingly view themselves and others by birthright as co-inhabitants and co-owners of Earth. This should then also come with certain rights and responsibilities.[10]

All nations have agreed to reach net-zero greenhouse gas emissions this century. Coal power is in free-fall toward zero. The wealthiest countries commit to become net zero in 2050 or earlier. China and India commit to 2060. And nations also aim to ramp up dietary changes and regenerative agriculture practices to improve the health of people and soils. Agricultural land has stopped expanding land use by 2030 globally. Hence, deforestation is halted, and locally adapted reforestation is growing everywhere.

Our four girls, Shu, Samiha, Ayotola, and Carla, now playful ten-year-olds, are growing up in a rapidly transforming world. They still have to deal with air pollution, extreme heat, floods, and fires. But by the end of the decade, Samiha and Ayotola have moved into new apartments with a school and hospital close by. Ayotola is excelling in math at school. As part of the universal basic dividend, Carla's parents receive an annual check of between $1,000 and $2,000 they can spend on anything. They save it for Carla's education. Over in Changsha, China, pollution around Shu's school is declining as the government encourages more micro mobility, such as e-bikes, mass transit, and electric vehicles, and renewables are phasing out fossil fuels fast.

Scenario Giant Leap: 2030–2050

The world achieves an end to extreme poverty (less than 2% living on $1.90 per day) in the early 2030s as Asia, Africa, and Latin America develop rapidly. A key foundation of economic growth in these countries is clean green energy (see chapter 7). Education systems are revitalized to promote local knowledge, culture, experience, and languages to "decolonize the mind," as Kenyan author Ngũgĩ wa Thiong'o once put it, alongside critical thinking and complex systems thinking.[11]

Policies to redistribute wealth fairly are now adopted in most countries and quickly ramp up. By 2050, the top 10% take less than 40% of the national incomes in all regions. In the US, Carla's family receives around $20,000 every year from the Citizens Fund, as a universal basic dividend. The fund is built from fees charged on top wealth holders, real estate, and on companies that use natural resources and public commons such as social media data usage. This means her family can afford to eat healthier foods, have access to better healthcare, and put money aside for education, hobbies, and travel.

Inequality, in disposable incomes within countries, is finally falling across the world, and as a result, trust is building. Governments have a stronger mandate to implement long-term policies on energy, agriculture, health, and education. Longevity increases globally. But since birth rates drop, population growth slows dramatically and peaks around 2050, well below the level in the Too Little Too Late scenario.

Around the world, people are eating healthier diets with less grain-fed red meat and more fruits, vegetables, nuts, and seeds. Food loss and waste is reduced along the whole food chain by better logistics, smart apps, and packaging. By 2050, most farms use regenerative and/or sustainable intensification techniques. Thanks to large public-private initiatives, trees and forests are starting to grow back on previously degraded lands, stopping the decline in the world's forests.

Greenhouse gas emissions fall precipitously during the 2030s and 40s, and by the 2050s, temperatures look likely to stabilize at well below 2°C, in line with the Paris Agreement. Nations must still contend with climate chaos, extreme heat events, wildfires, and rapidly rising sea levels, but governance systems are more resilient to cope with these shocks.

Scenario Giant Leap: Beyond 2050

Population peaks at around 8.5 billion and begins to fall in the last part of the century to reach around six billion by 2100, around the same level as in 2000. Along with renewable energy, regenerative agriculture, and healthier diets, this cuts overconsumption and material footprints, particularly among the richest 10%. This takes a lot of pressure off natural resources. Greenhouse gas emissions in the 2050s are now about 90% lower than they were in 2020 and still falling. All remaining atmospheric emissions of greenhouse gases from industrial processes are now removed through carbon capture and storage. As the century progresses, more carbon is captured than emitted, keeping alive the prospect of returning the global temperature to about 1.5°C above preindustrial levels. There is rising hope that much global biodiversity will thrive once again.

Shu, Samiha, Ayotola, and Carla are now thirty years old. They have all finished university degrees and are at the early stages of their careers. They do not expect to have the same career throughout their lives. Instead, they see opportunities to have several careers in different sectors. They'll retrain when they need to or want to, supported by an active state. Every month they receive a universal basic dividend. This provides a level of economic security that allows them to take more risks.

Thanks to a government relocation program and her universal basic dividend, Ayotola and her parents were able to move away from Lagos, threatened by the floods and rising waters. She works as an accountant, specializing in wellbeing indicators, and has decided to have one child. In Seattle, Carla has trained as an architect and designs passive homes for community housing, while her partner is a corruption analyst.

Samiha is now a food engineer, developing saltwater-resistant grains to increase yields. In her free time, she tutors children at the community center. Shu has chosen not to have kids, expending her time on social networks, and busy with the marketing and managing of huge fleets of shared electric carpools. Floods and storms are a regular occurrence, but measures have been taken to mitigate their effects—green spaces and trees in strategic locations—and new urban and sewage infrastructure make the city livable.

None of the women have ever experienced a stable climate, as once existed back in the 1900s. Extreme weather events come and go without toppling entire cities or nations. But they still cope pretty well because their governments are investing in their futures.

After reaching a peak during the 2020s, the Social Tension Index is steadily declining, indicating less social unrest. With improving wellbeing levels, citizens rediscover trust in governments, which in turn enables effective spending on education and universal health coverage. These factors feed back into more wellbeing and trust. A virtuous self-reinforcing cycle. Healthy diets are the norm. Universal healthcare is available to all. This leads to increasingly thriving societies that are resilient to shocks.

The Giant Leap scenario delivers way more on the Sustainable Development Goals than Too Little Too Late, and it does so while bringing the world back within planetary boundaries. The world is far from a utopia, conflicts still erupt, climate disruptions are still causing shocks, and the long-term stability of Earth is still deeply uncertain. But much pain and suffering is minimized. Extreme poverty is all but eliminated, and the risk of runaway climate change is diminished.

Figure 2.3. The Giant Leap (GL) scenario quantified as four time graphs visualizing the global developments in this scenario from 1980 to 2100. The curves are all relative to 1980 and highlights the dynamics between them. The global population was 4.4 billion in 1980 and peaks at 8.5 billion in the 2050s, before population starts a slow decline to around 6 billion in 2100. Income per person ("GDP per person k$/p/y") is 13% higher than in TLTL by 2050, and 21% higher in 2100. Notice that the net GHG emissions per person hit zero by 2050, on chart 3. Chart 4 shows different aspects of wellbeing in GL. The global Average Wellbeing Index first declines during the early 2020s transformations, but then improves dramatically for the rest of the century, as the impacts from turnarounds kick in and improve the prospects for long-term progress. Source: E4A-global-220501. The model and all data are downloadable from earth4all.life.

1. Main trends
Giant Leap scenario

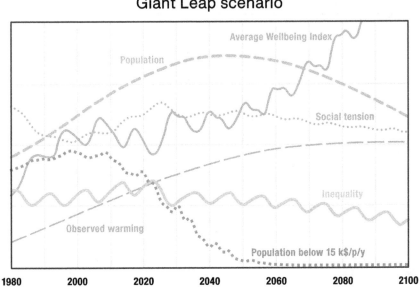

2. Human footprint
Giant Leap scenario

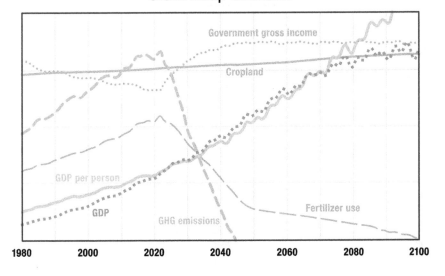

3. Consumption
Giant Leap scenario

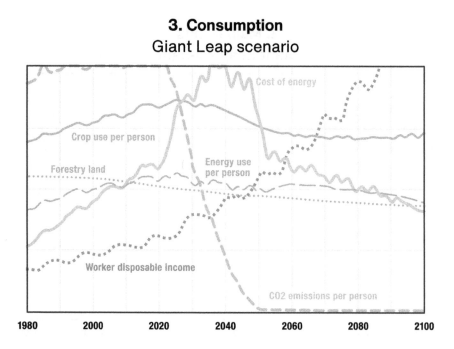

Cost of energy

Crop use per person

Forestry land

Energy use per person

Worker disposable income

CO2 emissions per person

1980 2000 2020 2040 2060 2080 2100

4. Wellbeing
Giant Leap scenario

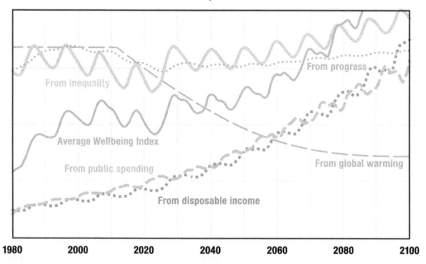

From inequality

From progress

Average Wellbeing Index

From public spending

From global warming

From disposable income

1980 2000 2020 2040 2060 2080 2100

In 2100, Shu, Samiha, Carla, and Ayotola celebrate their eightieth birthdays, and they look back at eventful lives. Shu reflects on how China solved its river pollution challenges. She is still amazed to see river dolphins alive and well in her city. Samiha's retirement is financed through the national pensions scheme and universal basic dividend, and she is writing a history of Bangladeshi women's rights. Carla lives in a passive home she designed herself, while Ayotola serves on the advisory board of the multitrillion Nigeria Citizens Fund.

Which Scenario Do We Co-create?

Now that we have sketched out the two main scenarios, we will look at the five extraordinary turnarounds needed to achieve the Giant Leap scenario. Why should people and politicians strive to make them happen, and how can they be realized? We present these extraordinary turnarounds in order: poverty, inequality, empowerment of women, food, and energy. In other words, addressing poverty accelerates movement on inequality, and progress on these two fronts helps accelerate the empowerment, food, and energy turnarounds.

The reason we start with poverty is that global poverty is still one of the very worst problems that humanity faces today. It has the greatest importance for Most of the World, where billions today live below a dignified living income. The poorest in the world suffer malnourishment, have greater health problems, and lack easy access to education and light at night (a significant barrier to a good education). They also lack a voice in the public debate. Can we turn this giant problem around much quicker than in previous decades in order to lift the quality of life for Most of the World? If so—exactly how?

How Good Is the Earth4All Model?

When we look back at some of the twelve scenarios presented in *The Limits to Growth* in 1972, we can see they tracked actual and important global trajectories in the last fifty years with surprising accuracy. We certainly hope that the Earth4All model will be equally successful, given the huge progress in understanding and data availability over the last fifty years. Yet, there are many reasons why some caution is warranted.

The Earth4All model, like all other simulations, cannot really predict the future. The model can only tell what are the consequences of the assumptions that are built into the model. Naturally we hope that we have included the most important assumptions, and in the right manner, but the Earth4All model is still a gross simplification of the real world.

So, while we believe the model is capable of exploring trends in important system variables, like fertility, economic growth, population, wellbeing, and climate change, we stress that the model is not at all capable of predicting the timing of future events and absolute values with any accuracy. The model is most useful in ensuring consistent thinking, for example, how investment in public services and economic development affects population and climate. Also, the model allows us to show the relative effect of policies, for example, how quickly poverty is eliminated in one scenario compared with another (poverty is eliminated one generation earlier in the Giant Leap compared with Too Little Too Late) or how inequality and social tensions evolve in each scenario: falling in the Giant Leap scenario, and rising in Too Little Too Late.

There are other reasons for caution in interpreting the model's conclusions. The world is moving into a less stable and therefore less predictable future. Geopolitical tensions are rising, countries are questioning the benefits of globalization, and we are seeing once strong democracies starting to decay. We have moved into totally unknown territory on the climate front. We have passed one threshold already: 1°C above preindustrial temperatures. Earth has not been this hot for the entire 10,000-year period of human civilization. We may expect to cross more thresholds this century, depending on decisions made in the coming decade. Food and energy production will exceed anything ever seen on the planet, with impacts on Earth's biosphere that will be huge, undoubtedly bringing many surprises. How will societies adapt to crowding, megadroughts, and extreme flooding? How will people contend with exorbitant energy costs and multiple breadbasket failures?

Our scenarios show quite devastating consequences for many regions around the globe; but even so, we cannot rule out the possibility that our scenarios are too optimistic given the uncertainties. The model can easily generate more gloom-and-doom scenarios if anybody feels the need for it. The Earth for All team did not. We chose to focus on the Giant Leap scenario, which builds resilience to shocks through greater societal cohesion and illustrates a possible way forward that leads to increased wellbeing rather than collapse.

3

Saying Goodbye to Poverty

Picture this: A woman in India throws a desperate look over her parched paddy fields. As a farmer, she is again reeling under the onslaught of another drought, this time worse than the last. Her income has been slashed. What rice she managed to harvest, she was forced to sell too cheaply due to undercutting by an international agrocompany. Now she can't afford new drought-tolerant rice seeds. Austerity measures in place mean that her state and federal governments are of little help. All extra state funds are funneled to repay debt obligations taken on during the last economic crisis. Climate change, poverty, and the failure of institutions combine to lock her—and her neighbors—into despair. Without seeds, what are they supposed to do now?

All low-income countries want to prosper and develop sustainably. But can they avoid the strategic pitfalls of Europe, the United States, Japan, China, and South Korea and instead develop in a fairer, cleaner way?

Our foresight analysis, building on historical cases, indicates it can be done. Rapid economic growth that is fair and clean is feasible but requires a new economic model. Under the current international structure, the policy options available to these countries are severely restricted and must be expanded. This will require a transformation of current global financial systems, trade agreements, and technology-sharing mechanisms. Crucially, there is an urgent need to remove the constraints on low- and middle-income countries to tackle the dual challenges of climate change and poverty. Without urgent action, it is exceedingly difficult for them to prosper economically and simultaneously reduce carbon emissions or adopt green technology.

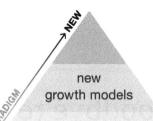

Figure 3.1. The poverty turnaround. Dramatic expansion in policy space forms the base of this turnaround. Along with massive changes in global finance, trade, and technology-sharing, this leads to a new economic model that allows the growth in low-income nations to rapidly reduce poverty in a green and fair way when combined with the following turnarounds.

The current carbon emission levels in the atmosphere are a by-product of rapid industrialization in the high-income countries over the last 150 years. However, today it is the low-income countries, or Most of the World, that have been left to deal with the consequences. These countries have an acute lack of resources and technology. Yet they are much more exposed to the climate emergency as a result of their geographical location.

We also know that the billion richest individuals account for 72% of the consumption of overall global resources, while the poorest 1.2 billion (the vast majority of whom reside in Most of the World) consume only 1%. Hence, the richest global societies are consuming the most natural resources while facing the fewest consequences, a deeply unfair situation.[1] The moral and historical impetus falls on the high-income countries to provide all possible support to low-income economies. This would constitute a central tenet of climate justice. Lifting hundreds of millions of people out of "extreme poverty"— where they live on less than $1.90 per day—would drive a global increase in emissions of less than 1%, according to new research.[2] That uptick could be countered elsewhere.

Poverty combined with climate change is not limited to low-income countries alone. Research has shown, for example, that poor and minority communities in the United States are worse affected by extreme weather phenomena like Hurricane Katrina.[3] The next chapter will focus on inequality within countries. This chapter will deal with inequality between countries: the challenges that poorer countries face and viable solutions.

What Is Our Current Problem?

Extreme poverty has declined dramatically in the last fifty years. But still almost half the world lives in poverty, surviving on less than $4 per day (equal to a GDP per person of some $1,500 per year). Pre-pandemic estimates on meeting global targets such as the Sustainable Development Goal 1 (eradicating extreme poverty) required low-income countries to grow at an average annual rate of 6%, and the consumption (or income) of the bottom forty countries needed to grow 2% faster than the average. COVID-19, however, is estimated to have set back progress on poverty by six or seven years. New economic estimates, accounting for COVID-19, indicate that up to 600 million could be living in extreme poverty by 2030 if economic development goes back to "business as usual."[4] Compounding this, the current economic system places low- and middle-income countries in a position where they must choose between addressing poverty and addressing climate change.

Issue 1: A Shrunken Policy Space

The policy space for governments seeking to act on both poverty and global warming is severely restricted by the global economic system. The free flow of finance is blocked by heavy debt-repayment restrictions. Through institutions like the International Monetary Fund or the World Bank, rich countries exert strong control over the finances of low-income countries, and extract considerable interest payments from them, leaving them with limited funds to invest in their own country. And foreign investors often extract more capital (when accounting for human and natural capital) than they interject. In the end, policies originally conceived to reduce poverty have either failed or, worse, exacerbated it.

Low-income countries lack funds (and savings) to invest in key development projects or infrastructure, such as power grids, water supply, roads, rails, and hospitals. More of these investments could spur a healthy growth model. Lack of such infrastructure, particularly electricity, causes African countries to lose up to 3 to 4% of growth per year, notes Masse Lô, founder and CEO, Institute of Leadership for Development, in an Earth for All Deep Dive paper.[5]

In many low-income countries, foreign investment is seen as a key solution. But the current global system encourages markets, rather than governments, to allocate the necessary funds or liquidity (cash) to these countries. To add insult to injury, these funds are in foreign currencies that means scarce financial resources being allocated to servicing ever-growing mountains of debt. This includes finance coming from other countries or multinational corporations.

This attitude is actively fostered by the global economic bodies. Governments are encouraged to have their economies completely open to capital flows. Highly mobile international capital can park itself in a country or sector it sees as profitable, but there is no guarantee that such funds will be invested either in poverty alleviation or in building energy-efficient capacity. Usually, this liquidity ends up in the financial sector of the economy, among stocks and derivatives, which are potentially quick to withdraw.

Foreign capital has modest to little impact on economic development, growth, or welfare in many countries.[6] Further, foreign capital often displaces (or crowds out) domestic investment or drives up greenhouse gas emissions and pollution.[7] Footloose finance, which is essentially looking to make a quick profit, is unlikely to be invested in the long-term projects that would address the development needs or clean energy capacity of a country. The investments for those purposes have to be intentional, coordinated, and strategic—goals that domestic economic policy stands a much better chance of meeting.

For some low-income countries, the reduced maneuverability from global structures is worsened by a large part of available resources being directed toward debt and interest payments. In 2020, debt in low- and middle-income countries rose to $8.7 trillion according to the World Bank. Of this, the debt burden of the world's low-income countries rose 12% to a record $860 billion. Most of this was related to pandemic needs.[8]

According to economist Richard Wolff, thirty-four of the poorest countries now pay far more in debt repayment (chiefly to rich countries) than they spend on the climate crisis. Similarly, the spending on healthcare during the pandemic has suffered. And as the debt burden rocketed in many low-income countries, their economic growth was stalling. Other countries, in seeking to stick to traditional recommendations by multilateral institutions, may maintain a very low debt burden. But they are only able to do so by limiting their expenditure on welfare schemes or on large capital-intensive green investments.

Issue 2: Destructive Trade Architecture

The expansion of global trade has naturally led to concerns about how much carbon dioxide is emitted at the various stages of production, transportation, and consumption of goods and services. Supporters of the current free trade model[9] argue that the current global trade architecture is compatible with poverty and climate goals under the right circumstances. Just shift trade to favor cleaner producing countries, and use it to motivate polluting countries to adopt technological solutions.

This can only happen, however, if high-income countries acknowledge the structural hurdles that prevent such actions from occurring. In actuality, the current global trade architecture hinders a shift to addressing both climate and poverty.

High-income countries outsource their production to low-income countries to benefit from reduced costs while low-income countries benefit from increased jobs and wages to their many workers.[10] But outsourcing has brought heavily polluting industries and more climate emissions to low-income countries. When assigning responsibility, however, the current standard method for determining carbon emissions is based on emissions within a country's boundaries, notes Earth for All Transformational Economics Commissioner Jayati Ghosh and team.[11] There is no accountability for consumption emissions.

The outcome of this process has been that high-income countries have been able to exploit cross-border trade to effectively "export" emissions to producing countries. The latter have now been tasked with the responsibility of cleaning up these exported emissions, but under stiff global competition. However, when countries attempt to clean up—through national regulations, protectionist measures, or controls on the importation of recyclable waste, for instance—they are then unfairly criticized for opposing free trade. And oftentimes taken to court.

This lack of distinction between consumption-based emissions and production-based emissions not only allows high-income countries to bypass responsibility, it also places the burden of potential tariffs on low-income countries, without providing them the knowledge/technology or financial resources to measure and control emissions. For instance, carbon border taxes aim to control emissions by taxing imports on goods produced in high-emission countries. Again, this shifts the monetary benefit to the high-income consumers rather than the low-income producers that are filling their demand for material goods.

Issue 3: Hurdles to Technological Access

From advanced materials to renewable energy, new technologies are key to solving global warming. Practically every model of climate response envisages some role for both existing and future forms of technology to cut emissions and environmental degradation. Unfortunately, much of this green technology is inaccessible. This is not due to technological infeasibility. It happens because the frameworks for tech transfer do not allow low-income countries to use them. From restrictive intellectual property laws to prohibitively expensive access in hard currencies, low-income countries that desperately need such technology to green their operations, bring vaccines to their poor, or reduce expenses cannot access it. Countries in need, who have already constrained financial policy space, are then further squeezed and compelled either to accept bad terms of use or else forgo access to such technology.

Turning Poverty Around: Addressing the Challenges

Our first turnaround, therefore, aims to enable three to four billion people to move out of poverty (see figure 3.2). Upgrading and rebooting failing economic systems to focus on both quality and quantity of growth will enable low-income countries to reach at least $15,000 per capita per annum (or some $40 per person per day) before 2050. At this level, these countries can deliver on most of the social Sustainable Development Goals (SDGs) such as food, health, education, and clean water to all their citizens. This turnaround is designed to future-proof low-income economies as they forge economies focused on wellbeing, while the interlinked energy and food turnarounds are designed to make this possible within planetary boundaries.

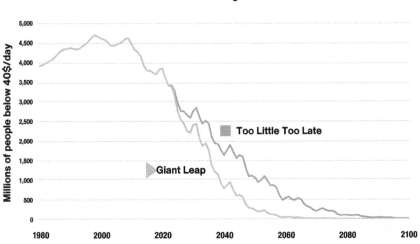

Figure 3.2. The end of poverty. The lines show how many million people live with an income per person below $40 per day ($15,000 per year). In the Giant Leap scenario, this goes to zero not long after 2050, while in Too Little Too Late, this happens a generation later, towards the end of the century.

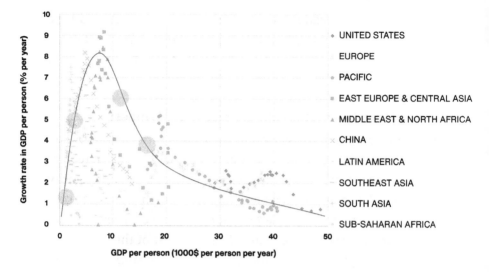

Figure 3.3. Economic development in all the world's main regions has followed a similar pattern: first the rate of growth rises to around 6–8% then falls to around 1–3%. Dots show regional growth rates for 1980 to 2020 as ten-year averages. The large circles show how the South Asia region can turn around from poverty toward mid-income levels by 2050, following a "growth guideline"—measured as incomes per person at constant purchasing power parity dollars (2017 PPP per person per year). Data sources: Penn World Tables 10; UNPD.

As countries get richer (see figure 3.3, horizontal axis), the annual growth rate (vertical axis) first rises from low values, and then peaks at around 6% to 8% before it starts a long decline. The poverty turnaround is achieved when low-income countries move from the left-hand side of the peak toward the right-hand side of the peak. To do this quickly, low-income countries must gain the policy space needed to act, as well as just and viable terms of access to necessary finance and technology.

Solution 1: Expand Policy Space and Deal with Debt

Building a productive infrastructure is hugely dependent on upfront investments. Such mobilization of resources is extremely unlikely to occur quickly enough without active steps undertaken by the governments of the respective countries. Fiscal policy—the choices and methods through which the government is able to spend—can be an effective tool in generating the required investments. Mobilizing these investments requires a combination of increased public invest-ment as well as changing regulations and incentives to ensure that private investment also aligns with these goals. Different means exist to help finance public investment—increasing taxes on the rich and on large corporations, deficit financing, and creative use of central and development banks are all viable options.

However, while these methods to expand the policy space are available in principle to low-income nations, in practice they are heavily discouraged by the larger international financial architecture. To reverse this, a first step would be to grant debt relief to low-income countries that require it. This relief will help these countries to com-mit more funds to protect against climate change and transform their economies. Greater global coordination to increase corporate taxa-tion would help relieve the pressure on smaller countries whose tax-based policy options are restricted.

Given the global nature of the challenge, cooperation on a Global Green New Deal could be a possible measure to shift the global pro-duction systems to greener paths as well as generating millions of high-paying jobs throughout the world, which could deal a blow to the bane of poverty.

Finally, high-income countries can legislate to regulate or other-wise restrict transnational corporations from investing in brown industries (highly polluting or carbon-intensive sectors) in low-income countries and instead channel their investments toward green industries (sectors that are less carbon-intensive or are supportive of the larger process of transitioning to a sustainable economy). Low-income countries seeking to grow their economies have little choice when large conglomerates seek to invest in polluting industries, and thus governments in high-income countries must necessarily take up a greater responsibility in regulating their own companies.

Solution 2: Transform the Financial Architecture

Debt is like deep snow on a steep mountainside. The smallest of move-ments can trigger an avalanche. Because low-income nations depend on foreign capital for reserves (liquidity in the international economy), capital flight is all the more damaging. When money rapidly flows out, low-income economies are unable to pay for crucial imports. When countries have high levels of debt, not only does capital flight reduce reserves but it also weakens the national currency, increasing existing debt and interest payments. The lack of liquidity, augmented by debt, can cripple a low-income country's economy. It leaves countries unable to generate any positive investment and reduces the ability to spend on climate and poverty alleviation.

During the pandemic, some countries, like the United States, had the incredible privilege of being able to inject trillions of dollars of cash into their flailing economies through government loans and "quantitative easing." Can other countries adopt similar solutions? In theory, any country with its own currency can. But in practice, it is trickier. However, since 1969, the International Monetary Fund (IMF) has been empowered to lend in a country's currency through Special Drawing Rights (SDRs). SDRs are an international reserve asset (that acts like a currency) that provides a backstop to maintain confidence in a country and its monetary system. The problem is that SDRs benefit rich nations far more than low-income countries.

In 2021, to counter the pandemic shock, $650 billion in SDRs were allocated for IMF members to use. Unfortunately, the quota-based system of the IMF means that SDRs are granted based largely on GDP, and so $400 billion of this went to rich countries that would be unlikely to need them. Even with this imperfect provision, the allocation helped several low-income countries struggling with balance-of-payment issues. Notably, SDRs do not add to countries' debt burdens and are nonconditional. There is thus ample scope to use SDRs more effectively—ensuring greater SDR availability to low-income nations, exploring the option of allocating SDRs more frequently, using SDRs as the basis for forming climate finance trusts, and channeling SDRs to regional development banks for use in climate-related investment efforts.[12] Over the medium term, the entire foreign-denominated debt and trading system needs complete transformation to enable countries in Most of the World to borrow at low cost in their own currencies. This simply means giving all countries the same extraordinary privilege accorded to high-income countries today.

Another promising solution could be a new multilateral institution: the International Currency Fund (ICF) to establish two-way markets in currencies, especially for longer durations for which no private market currently exists. Complementary to SDRs, the ICF could improve how financial markets work by finding and acting as a counterparty to investors, borrowers, donors, corporates, and remitters of foreign exchange to offset currency exposures, risks that occur when investments are in another country's currency. Its multilateral treatment as a preferential creditor would reduce collateral required for trades and allow it to offer more products that aid local market development, increase liquidity, and attract private investors to currency risk as an asset class.

Solution 3: Transform Global Trade

Reforming the system that guides global trade requires several important steps in order to address the barriers that affect low-income economies.

A most radical need is to reconsider how carbon dioxide emissions are accounted for in trade agreements—most importantly, carbon dioxide emissions calculations and the consequent policy controls need to distinguish between production and consumption of goods and services. This would mean not just looking at the total emissions within a country's boundary but instead identifying the source—whether from consumption of goods or production of goods—and treating them differently for taxation and regulation. This would ensure that countries that have historically played a minimal role in climate change are not unfairly punished for seeking out the same growth outcomes that other countries have enjoyed, and similarly that countries seeking to simply outsource their carbon dioxide emissions are unable to do so. This would also constrain material footprints where they need to be constrained.

In a similar manner, there is a need to revive the concept of the "infant industry model"—shielding new industries in a country from global competition—with import restrictions. This model was so successful in helping economies like South Korea and China escape the middle-income trap. By recognizing the need to protect green industries from competing prematurely with larger established international players, countries are more likely to develop local green investments that are sustainable in the long term.

Finally, there is a need to reconsider the role of regional trade. You would never see a light-weight boxer being forced to compete with a heavy-weight. Likewise, promoting and protecting trade across low-income countries makes a lot of sense. Additionally, encouraging the matching of regional production and demand can help shorten supply chains, build resilience around fledgling industries, and help ensure that newer green markets have the time to grow.

Solution 4: Improve Access to Technology and Leapfrogging

The hurdles that prevent low-income countries from accessing technological solutions to climate change and poverty must be similarly overcome. There are thankfully a number of steps, both in the immediate and in the medium/long-term, that are available.

In the immediate term, the existing intellectual property rights systems need to be used to mandate access to technology. International treaties related to intellectual property rights were initially designed to require intellectual property holders to give access to others. The World Trade Organization (WTO) Agreement on Trade-Related Aspects of Intellectual Property Rights (TRIPS) is one example. Over time, though, the WTO case law against low- and middle-income nations has heavily diluted these provisions. Expanding and strengthening them to target climate change could greatly speed up the process of technological transfer. Similarly, a TRIPS waiver—granting immunity from legal challenges in the WTO for domestic intellectual property rights laws—needs to be considered for green and health technologies that low-income countries require but cannot access.

Going a step further, countries from which such technology originates can legislate, incentivize, or otherwise compel corporations to enter into agreements with companies or governments in low-income countries. Some corporations deceptively use treaties to

avoid substantive technology transfers.[13] The governments of origi-
nating countries are much better positioned to regulate or compel
multilateral corporations than low-income countries and must therefore take greater responsibility in doing so.

Finally, there is an urgent need to overhaul the entire intellectual property rights regime and support the use of patents in a more
responsible manner. The selective application of patents can be useful as a short-term, targeted incentive for investment, but the current
system heavily encourages simply holding onto long-term patents
for the sake of licensing and other forms of revenue. Strategically
shifting toward a system that makes innovation more accessible can
help ensure that as new technologies are developed, they can be more
readily deployed to countries that require them.

Barriers to the Solutions

If the solutions are so clear, what is keeping low-income countries
back? Many of the proposed solutions are quite radical, and thus can
be expected to experience significant barriers to implementation.
Inertia and path dependence—the development of systems that rely
on choices made decades or even centuries ago—ensure that the current financial system is heavily in favor of maintaining the status quo.
From the pervasive free-market model of growth to the heavy anti-
regulatory bias that has emerged in contemporary policy, there are
major roadblocks holding progress back.

Multilateral Institutions

In order for low-income countries to attract financial capital, multilateral institutions like the IMF and World Bank have pushed them to
undergo several "disciplinary" or "macroprudential" reforms. These
reforms tend to privilege the perceptions of private cross-border
investors, rather than the interests of citizens. This discourages state
intervention to address issues such as poverty and basic needs. They
have forced states to curb their ability to use debt to finance welfare
programs, placing restrictions on capital flows or even raising taxes.

Countries that do not adopt such reforms, or that deviate from them, are downgraded by international credit-rating agencies and seen as risky investments.

These credit ratings are indeed a decisive factor in how much a government is able to help its citizens during a crisis. A recent research paper has shown that credit ratings are correlated with the size and promptness of fiscal relief that countries were able to provide during the COVID-19 pandemic.[14] Governments of low-income countries are extremely wary of losing credit scores since a downgrading could result in capital flight, or funds being moved out of the country, leading to an economic crisis. To retain their ratings, they even engage in harsh austerity measures and limit essential public investments required for poverty alleviation and a green transition.

Perceived and Actual Corruption

The conventional concern raised when considering the policy options available to low-income countries is that corruption—both perceived and actual—must be addressed before high-income countries and their corporations can reasonably and reliably engage in investments. Certainly, concerns about how institutions operate may present a barrier to any systemic change. An additional important consideration, however, is the role that such corporations themselves play in ensuring that the barrier of corruption remains erect. In South Africa, for example, the Zondo Commission's reports into allegations of fraud details the role of transnational corporations in initiating and fueling massive corruption in the country.[15] Firms seeking to minimize costs may find it easier to simply exploit the laxity of legal institutions, and this other side of the equation may also present a large barrier.

Arbitration and Litigation

The current legal framework is heavily skewed in favor of high-income countries and transnational corporations. Countries like India and China have faced the legal might of WTO when seeking to expand production of solar panels. Similarly, large companies like Monsanto have zealously used the legal system to pursue farmers in low- and middle-income countries over seed-related patenting violations. The argument—that protection of such a system is necessary to encourage innovation and investments by private players—ignores the reality that the regime ends up locking out countries and groups who require the technology to shift away from polluting practices and must then either abandon technological change or divert much-needed funding from other areas. Where larger private corporations are concerned, this is usually done with the backing of their respective governments. The ultimate outcome is a reduction in the capacity for low-income countries to create green jobs and build productive green capacity.

Conclusions: The Poverty Turnaround

Four essential actions for achieving the turnaround in low-income countries have been provided: expand their policy options, address the impact of debt and the larger financial infrastructure, reconstruct the global trade architecture, and fix the systems of technology transfer. These are not the only actions needed, but they are certainly a

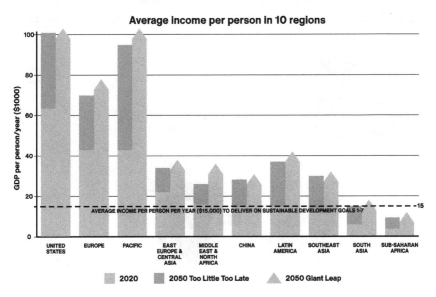

Figure 3.4. Incomes per person (GDPpp $1,000 per year): in 2020; in 2050 with Too Little Too Late; and in 2050 with Giant Leap. Source: E4A-regional-220401.

necessity to ensure that progress in fighting poverty in a climate-conscious manner can happen. Figure 3.4. visualizes the results per region, showing the incomes per person in 2050 without the turn-around (as in Too Little Too Late, TLTL) and if the turnaround is completed (as in Giant Leap, GL).

In sum, this amounts to a call for active and competent governments capable of increasing the wellbeing of their populations. A key will be to increase labor productivity in the early decades of economic take-off, and acknowledge that market solutions will not suffice if the goal is to build a strong nation for the benefit of its working majority.

The energy and food revolutions that will sweep the world this generation are a once-in-a-century opportunity for transformation, and the economic opportunity presented by such a comprehensive overhaul is immense for low-income countries. These economies can leapfrog over obsolete technologies. They can avoid appalling pollution, and this can help them escape some of the legacies of historical global inequality.

The scale of this transformation cannot be underestimated. Economic growth of at least 5% per year across low-income countries must be catalyzed immediately to achieve the SDGs by 2050. But

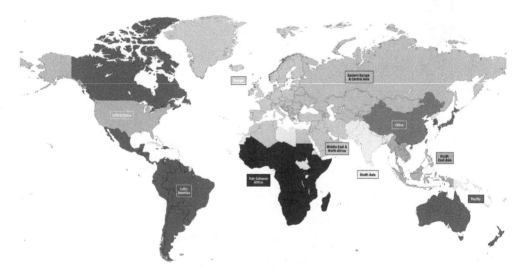

Figure 3.5. The ten world regions as used in Earth4All. Source: mapchart.net.

many low-income countries have seen growth stagnate in the last decade. The pandemic has further crippled these vulnerable economies, and adapting to climate change is beginning to drain scarce resources in low-income countries. This is an emergency situation.

This first turnaround has focused on low-income countries, but of course, many of the solutions will apply to middle-income and high-income countries too. The global financial system, after all, was built in a bygone era. It has served a purpose and has contributed to some forms of peace, stability, and prosperity; but its fault lines are visible everywhere, and it is clearly not fit for purpose in the Anthropocene. Ultimately, though, building a new future-proof economic ecosystem will require a shift away from a myopic focus on quantity of economic growth. The new ecosystem will only succeed if it prioritizes well-being—that is the quality of economic growth. That requires dealing, next, with rampant inequality.

4

The Inequality Turnaround
"Sharing the Dividends"

Countries where citizens are economically more equal function better. They have greater social cohesion, an essential ingredient for making long-term collective decisions that benefit the many, not the few. So, if our Earth for All goal is collective action to protect civilization in the long term, then greater equality is as close to a silver bullet as we might find.

New data has allowed us to see an unambiguous pattern related to income inequality over the last decades. Countries that are more equal perform better in all areas of human wellbeing and achievement than countries with divisive levels of income inequality. Whether low-income (such as Costa Rica) or high-income (such as Scandinavia), more equal countries tend to have better outcomes when it comes to trust, education, social mobility, longevity, health, obesity, child mortality, mental health, crime, homicides, and drug misuse. The list goes on. Counter-intuitively, even wealthy people in more equal societies, for example the Nordic countries, have higher levels of wellbeing than wealthy people in countries with high inequality such as the United States, Brazil, or South Africa.

The equality turnaround consists of three main levers that steadily push toward a new economic paradigm:

- More progressive taxation on both income and wealth for individuals and corporations.[1]
- Strengthening labor rights and trade unions' negotiating power.
- Safety nets and innovation nets to share prosperity and provide security during a period of deep transformation, for example, big ideas like universal basic income/dividend programs.

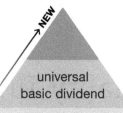

Figure 4.1. The inequality turnaround. Progressive taxation and fees on wealth distributes incomes more fairly; workers gain protections and fair compensation through re-unionization and other mechanisms to empower workers; universal basic dividends allow citizens to benefit from shared resources in a Citizen Fund. This has co-benefits including providing essential safety nets for citizens during these transformative decades.

Throughout this chapter, we will focus on economic inequality *within* countries—the differences in income and wealth between rich and poor, whether in low- or high-income countries. Inequality is broader than just wealth and income, but other aspects of inequality, for example gender equity, are tackled elsewhere (see chapter 5 on empowerment).

The major challenge is that within-country distribution has recently been moving the wrong way. Decade by decade, countries have become more unequal in every region of the world with the exception of Europe. The poorest 50% of people take less than 15% in total earnings, while the richest 10% take well over 40%. In many regions, it is closer to 60%.[2]

A key goal of the inequality turnaround is to ensure the richest 10% in societies take no more than the total income of the poorest 40% in society. This means that four poor people together have the same income as one person in the top 10% per year. This is seen as a tolerable level of inequality.[3] Above this level of inequality, social and health problems get much more severe and there is less social cohesion, making it more difficult for governments to make long-term decisions.

Our Earth4All model tracks changes in both wellbeing and inequality, and then captures this through a novel "Social Tension Index" and an "Average Wellbeing Index." The Social Tension Index is an indicator of potential polarization particularly in relation to inequality. A rising index indicates strengthening polarization, which is often not conducive to strong collective action within and across societies. The scenario Too Little Too Late sees social tensions rise as the wealthy elites become increasingly powerful and distant from everyone else. The model cannot predict what will happen next, of course, but it is easy to imagine scenarios where political institutions struggle to bring all on board a journey of deep transformation. Boom and bust economic cycles and shrinking safety nets for the most vulnerable drive despair and resentment. Earth for All Transformational Economics Commissioners Richard Wilkinson and Kate Pickett make the major point[4] that if countries implement a series of policies to share the wealth and commit to greater equality, this opens the door for the psychological conditions that build the level of social trust necessary for any transformation to sustainability, such as in the Giant Leap scenario.

How can we achieve this level of equality? Let's clarify the three broad groups of solutions to reduce inequality in some more detail before we review the problems that must be overcome:

More fair distribution of disposable incomes is the starting point. This can be achieved through *progressive income taxation*, meaning a higher tax rate for those with higher incomes. But this is not even half the story. Inheritance and wealth taxes also need to become more progressive because financial assets are accumulating at far greater rates than incomes are increasing. This means the gap between rich and poor will inevitably get wider and wider until, that is, wealth accumulation is taxed at a proportional rate. Globalization means that we need greater international efforts to close financial loopholes and stem the flow of money into offshore tax havens. This means stronger oversight of transnational corporations. A minimum international tax rate on businesses can also support greater equality. This was agreed in 2021 by most wealthy nations, to the surprise of many—but at a paltry 15%.

Labor rights and the *negotiating strength of workers* must be enhanced in order to increase the worker share of national income. Collective bargaining needs to be greatly supported after decades of erosion of union and worker power. More workers should be empowered with co-ownership and seats in the boardroom to influence decisions. During a turbulent period of transformation for industrial sectors such as energy, food, transport, and heavy industry, such changes are essential to ensure that workers support bold company action and benefit from the economic turnarounds, rather than fall further behind.

Finally, our boldest proposal is to explore Citizens Funds that would pay out universal basic dividends in all countries (see more on this in chapter 8). This proposal is based on proven effective ways to transfer a portion of the wealth extracted from common resources such as fossil fuels, land, real estate, or social data. In addition to redistributing wealth more fairly, this will provide essential individual economic security during the transformation of societies, and it is likely to spur creativity, innovation, and entrepreneurship.

The Problems with Economic Inequality

Between 1950 and 1980, inequality within countries actually shrank in Europe, the United States, and a few other places. This remarkable achievement occurred during three decades of enormous social, political, and technological transformation and rising prosperity following the World War II. Since 1980, the gulf between the top wealth holders and the rest has grown ever wider decade by decade. Today, when it comes to wealth, half the population of the world owns a tiny 2% slice of the global pie, while over three quarters of the pie (76%) is grabbed by the richest 10%.[5]

A politically acceptable solution to inequality, you might think, is to grow the pie even more: economic growth will eventually take care of the problems, everyone's prosperity and wellbeing will spiral ever upward, and all will be happy eventually. For low-income countries, this can work to some extent: life expectancy, education, wellbeing, and happiness shoot up in early stages of a country's development. But among high-income countries, further strong economic growth no longer connects to improving health, wellbeing, or happiness. Some rich countries are almost twice as rich as others but show no sign that the populations enjoy better health or wellbeing.[6] The current massively financialized globalized economies ensure that, as the pie grows, those already holding the largest shares snatch an ever-greater slice.

Why is inequality so bad for societies? The argument goes that high inequality provides necessary incentives to work harder. But evidence does not support this. What evidence does show unequivocally is that extreme inequality is a destructive force in societies.

Skewed Political Power

Wealth is an important source of economic and political power. Extreme inequality means power is increasingly concentrated in the hands of the super-wealthy and most valuable companies. This is deeply destabilizing for societies in general and especially for democracies because a foundational principle of democracies is fair representation.

Take the catastrophic failure of the global financial system between 2007 and 2009. People in the financial sector wield extraordinary political power. To prevent complete meltdown, governments stepped in, finding trillions of dollars to shore up failing banks. Citizens across the world were forced to pay for this failure through harsh austerity measures. Badly run banks were saved and went on to make ever-greater profits—with the same owners. The sense of injustice that ensued is linked to a rise in populist leaders promoting division and misinformation within societies.

Another example is the Gilet Jaune, or Yellow Vest movement in France. Following the Paris Agreement on climate in 2015, French president Emmanuel Macron proposed a fuel tax to encourage a shift away from cars and vans with high greenhouse gas emissions. Petrol prices rose, sparking many weeks of protest that birthed a movement eventually involving millions of people. And so, a long-term policy to reduce emissions was undermined by low-paid workers and middle-class workers hit hardest by decades of economic stagnation. The protesters were convinced that the political system is in the palm of the wealthy elite and refused to play ball. Solutions should be acceptable to the majority or they risk catastrophic failure leading to potentially decades of incremental progress.

Overconsumption Among the Rich

Besides political representation or lack thereof, inequality has another pernicious effect on societies. In extremely unequal societies, the desire for higher status drives extreme materialism and luxury carbon consumption.[7] Globally, nearly half (48%) of all emissions are commandeered by the richest 10%. And the richest 1% are responsible for a phenomenal 15% of all fossil fuel emissions on Earth and growing rapidly.[8]

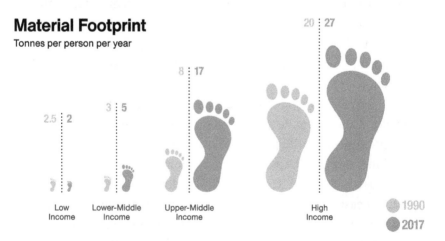

Material Footprint
Tonnes per person per year

| | | | | |
| 2.5 : 2 | 3 : 5 | 8 : 17 | 20 : 27 | |

Low Income   Lower-Middle Income   Upper-Middle Income   High Income   ● 1990 ● 2017

Figure 4.2. Between 1990 and 2017, material footprints—the material consumption per person measured in metric tons per year—grew dramatically for upper-income and high-income consumers while they shrank slightly among low-income consumers. Data source: UNEP & IRP *GRO-2019*, fig. 2.25.

In very unequal societies, people are more likely to feel anxiety about their status in society, worry about how other people may judge them, and hunt for designer labels, more expensive cars, and other products that signal status. The SUVs and endless darting between continents come with a high carbon price tag. But this is a zero-sum game on an infinite loop—*we can't all have higher status than everyone else*. And the pressure to consume can make it all come crashing down. Debt and bankruptcy are more prevalent in more unequal societies.

Tackling the drivers of excessive material consumption, and focusing our systems on what people fundamentally need, will contribute to accelerating all five of Earth for All's extraordinary turnarounds, notes Earth for All Transformational Economics Commissioners Anders Wijkman and Lewis Atkenji. We know that overconsumption of natural resources drives ever-increasing environmental impacts. It also brings negative consequences for wellbeing. The health impacts of food overconsumption, for instance, are well-known. Furthermore, mental health issues are concentrated in countries where resource consumption is high. As Earth for All's turnarounds recognize, inequality reduction will be crucial for a sustainable future: Those in the wealthiest countries are disproportionately responsible for the world's environmental impacts. Much of elite consumption places excessive costs on the rest of society—costs that are not covered in the market price of the goods consumed. Correcting this requires fundamentally changing how we govern our societies and economies.[9]

Enclosure of the Commons = Privatization

One driver of wealth accumulation, and so inequality, was put in place centuries ago: enclosure. Much of our land was once managed as a commons—collectively shared and sustained. But over time a new management system evolved. Governments, colonial powers, or other "authorities" granted property rights and ownership of land. Little by little, once open lands, common to all and managed sustainably for generations, became enclosed. That is, privately owned for the sole benefit of the owner.

Eventually this system of management entirely dominated. The use, access, and benefits of a resource became tightly controlled by the "owner." This expanded to other resources: minerals, data, and knowledge through patents. But "ownership" is never as clear as it seems. Government-funded research, paid for by taxes, led to many of the key technologies within mobile phones. The Internet, GPS, touch screen, even Siri, Apple's AI assistant, started life at universities. Perhaps a wider group of people, or even all citizens, should share in the ongoing wealth generated by enclosure of land, data, and publicly funded knowledge?

Even though these natural or intellectual resources may be considered part of the collective wealth of a nation, the accumulating riches derived from these resources are often handed down through affluent families across the generations, often through mechanisms that avoid taxation. Enclosure means wealth accumulates at a rate that is higher than the growth in income from labor, so, naturally, wealth will outpace income—opening up an ever-wider chasm between rich and poor.

Warning: Disruption Ahead

Whichever path the world follows, from Too Little Too Late to Giant Leap, and everything in between, the coming decades will be disruptive for too many people, in large part due to the legacy of chronic inequality.

Exponential technologies, from solar power to genomics, will continue to disrupt our lives, and this disruption is likely to accelerate. Innovations are moving from the dreams of entrepreneurs and engineers into the lives of billions of people. Accelerated artificial intelligence, machine learning, amazing mobile Internet speeds, and robots will increasingly take over tasks at work that once kept people busy. The advantage is that this will free them from drudgery and make them available for green jobs, employment in the caring professions as societies deal with aging populations, or the knowledge economy. And this can work well—but only as long as some of the super-profits in the robotized sectors are used by governments to invest in retraining, and new jobs in an economy that builds well-being for the majority.

This technological disruption will increase inequality in society, often in ways that are difficult to predict, like the way social media has at once connected more people but also industrialized misinformation and so helped undermine democracy. In parallel to social media disruption, technology, from robotics to the Internet, has exerted a downward pressure on wages, creating a class of gig workers and the so-called precariat that live day to day on zero-hour contracts in "fulfillment centers" where their movements are driven by algorithms.

Technological disruption is just part of the picture. China's rising economic and political power will change the geopolitical order. India will likely become the most populous country on Earth, and its economy may grow rapidly if properly managed. We can expect further disruption from climate change and other environmental hazards as we edge toward the 1.5°C threshold. And, of course, we could be hit by a black swan—a massive event that could not be predicted. We could even be hit by a flock of black swans. Or by other high-impact events that can be predicted and prepared for in advance but normally ignored by voters and their politicians, for example, new and deadlier pandemic risks.

The point is: Societies need to prepare for disruption and build for resilience—this means providing essential safety nets. One important way to build resilience is to tackle inequality.

Measuring Inequality

A starting point for addressing inequality is to measure it. For a century, the most common way to do this has been to calculate the "Gini coefficient" of a country, an index named after its creator, the statistician and demographer Corrado Gini. This index measures income distribution from poorest to richest in a society. But due to several drawbacks, including its complexity, not everyone likes the Gini coefficient. More recently, economist Jose Gabriel Palma has argued that what really matters is how much income or wealth goes to the richest 10% compared to how much goes to the poorest 40%. His argument is pretty sound: Statistics show that the middle 50% takes about half of gross national income irrespective of country or time; so what really matters is what is happening toward the extremes.[10]

The Palma ratio, therefore, is quite simply the share of total in-

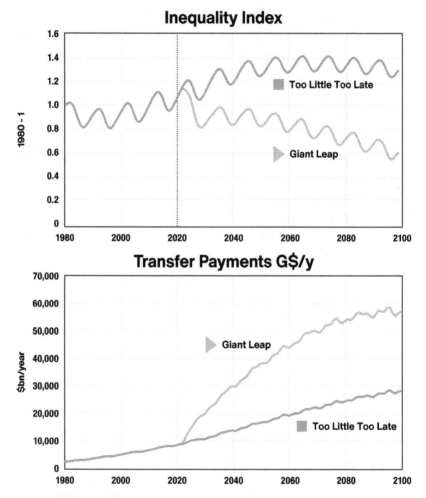

Figure 4.3. Inequality turnaround: The lower Inequality Index in Giant Leap is due to higher transfer payments (from taxes and Citizens Fund dividends) than in the Too Little Too Late scenario.

come captured by the richest 10% divided by the share taken by the poorest 40%. Scandinavian countries have a Palma ratio of about 1.0. This means the richest 10% take about the same total income as the poorest 40%. In the United Kingdom it is 2.0, in the United States it is 3.0, and South Africa has a Palma ratio of 7.0. We argue a ratio of 1.0 is a sustainable level of inequality. We can show that, over long periods, a 1.0 ratio maintains strong social cohesion and supports very high levels of wellbeing for the majority.

In Too Little Too Late, inequality continues to rise across regions. But in the Giant Leap scenario, inequality is significantly lower, as more progressive taxation and bigger transfers from rich to poor make workers' disposable incomes go higher than in Too Little Too Late.

Social Tension Index

One innovation in the Earth4All model is a Social Tension Index. It measures the perceived rate of societal progress, that is, increase in wellbeing. The index is linked to inequality because nations with greater inequality are less able to govern effectively. The index rises

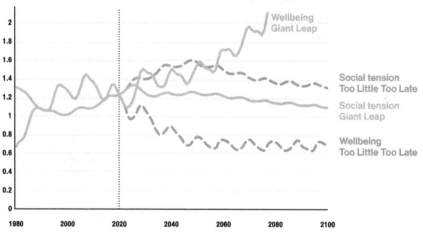

Figure 4.4. Social tensions run higher in Too Little Too Late (*dashed lines*) compared with Giant Leap (*solid lines*) toward 2100. Source: E4A-global -220501.

significantly higher in the Too Little Too Late scenario compared with the Giant Leap.

A Giant Leap Toward Greater Equality

The Giant Leap scenario succeeds by driving rapid shifts toward greater equality across all ten regions of the world out to 2050. This is driven by implementing three key solutions, each one at a greater scale of ambition: progressive taxation, including a wealth tax; empowerment of workers; and a fee and dividend approach to exploitation of natural resources. We argue all have a part to play in driving a Giant Leap, but regions and nations might apply these solutions differently. One size is unlikely to fit all, and we want to emphasize that this does not rule out other excellent solutions to reduce inequality. We will mention some of these as we go on.

Progressive Redistribution of Income and Wealth

Let's start with income. There are several ways to ensure incomes can be tamed to protect against destructive levels of inequality.

The main and obvious way to close the income gap is through progressive income taxation, with those on the lowest incomes paying little in tax (or even none if jobs are threatened by automation) while the highest earners are taxed more heavily.

A second challenge is to tame wealth accumulation. Inheritance taxes and wealth taxes must be increased to prevent wealth accumulating faster than workers' incomes—in percent per year. Without such an intervention, the gap between rich and poor will inevitably get wider and wider. A famous example of the scale of the problem was US billionaire Warren Buffett's incredulity over paying a lower tax rate than his secretary—because his income is coming from investments not a salary.[11]

A third solution, with a huge psychological impact, would be to place limits on how much the salaries of company executives can exceed the average salary in the organization. Executive-to-worker pay disparity has skyrocketed in recent decades. A 2021 study by the Economic Policy Institute found that America's largest public firms paid CEOs 352 times that of the average worker in 2020. In 1965, before the dramatic expansion of the wage gap, the CEO-to-worker pay ratio was 21-to-1.[12]

Globalization means that we need greater international efforts to close financial loopholes and stem the flow of money into offshore tax havens. Of course, transnational corporations must also be held accountable, and here we can report progress. In 2021, for the first time, the G20 group of high-income countries agreed on a foundation for an international corporation tax. This is a significant achievement to begin to address the race to the bottom.

Economic Democracy: Retraining and Empowering Workers

Over the last four decades, the negotiating strength of workers has been deliberately weakened to the point that even in the richest nations on Earth zero-hour contracts are now normal. The common argument for the erosion of the power of trade unions and workers more generally has been to improve competitiveness in an increasingly globalized cut-throat economic world.

In high-income countries, the manufacturing sectors shrunk considerably in recent decades as manufacturing moved to middle-income countries. In their place, the service sectors have grown, with lower wage rates because unionization has been aggressively resisted. A first priority is to help workers regain past strengths. Governments can also provide guaranteed jobs doing public work, paid for by higher taxes on the rich. The need is high for workers that can provide essential environmental and social services, including tree-planting, rewilding, and soil protection. If trade unions in these countries cannot gain their former strength, other solutions to improve economic democracy exist.

Many solutions have their roots in democratizing workplaces. Employee co-ownership plans can give workers a stake in the companies that employ them. Seats for workers on corporate boards would allow workers, shareholders, and company executives to take decisions collectively. More worker cooperatives can play key roles too. All these measures allow workers to benefit from the essential economic turnarounds, which in turn will drive worker support for bold plans rather than drive workers to resist change.

We also recognize that unpaid workers (largely but not only women) provide priceless services to economies and society, enhancing social cohesion. How can this transformative moment be used to not only acknowledge their contributions to society but protect them, reward them, and empower them? We'd like to introduce the idea of a universal basic dividend.

Introducing Citizen Funds and a Universal Basic Dividend

In recent years, several promising ideas to help redistribute wealth and normalize economic security have been proposed, trialled, and even successfully implemented. French economist Thomas Piketty proposes giving all young adults $100,000 to start their working lives with a healthy degree of economic security. A universal basic income has been trialled to limited extents in Finland, Canada, Ireland, and Kenya, among other places. There are many different models (and designing workable programs is not easy), but essentially all citizens receive a small regular income regardless of working status.

In the US since 1976, the Alaska Permanent Fund collects a share of revenues from oil companies extracting the state's natural resources and pays a dividend to *all* citizens. This typically ranges from $1,000 to $2,000 per person every year. For 2021, the dividend was $1,114, meaning that a family of four would receive $4,456. An extension of this idea was proposed in 2017 by Republicans associated with the Climate Leadership Council (CLC). The CLC proposal calls for fees linked to carbon emissions to be redistributed to all citizens across the United States. It received bipartisan kudos as one of many necessary solutions, but not as a rationale for relaxing or forgoing other carbon-related regulations, as its proponents have proposed. CLC estimates that a family of four would receive about $2,000 every year from such a system, creating some economic security during upheaval and transformation.[13] (We'll discuss dividends in greater detail in chapter 8.)

All these proposals have merit. They provide a level of economic security during transformation, they prevent workers being forced to accept rock-bottom wages, and they give workers the power to say no to exploitation. But they can also spur creativity, innovation, and entrepreneurship by creating economic freedom. This is more than just a safety net, it's an innovation net.

Building on these proposals, and acknowledging the transformations to come, and the risks and deep uncertainties that will accompany these changes, we propose providing universal basic income through a Citizens Fund. Companies that, for instance, produce carbon dioxide emissions, contribute to deforestation, use public data, or mine on land or in the deep sea would pay a fee for exploiting common resources. Governments would distribute this revenue equally among all citizens as a dividend.

You might ask, wouldn't it be fairer to just give the dividend to the least well-off in society? It is important to bring everyone in society on this journey or it risks failing, so the solution has to work for the majority. We know that if the middle classes feel they are benefiting from a policy, then they will support it. We know if the middle classes feel others are benefiting from their hard work, then they are less likely to support a policy. A universal basic dividend also has the benefit of simplicity. It is easy to communicate and this increases its chance of broad support.

We've discussed three broad groups of solutions to drive greater economic equality: progressive taxes on income and wealth, empowerment of workers, and a Citizens Fund or similarly bold fee and dividend initiative. There are plenty of other good solutions out there that we certainly support too. During economic meltdowns, central banks have ensured companies survive often by buying stocks at discount rates. What if governments held on to these stocks when economies recovered? In this way, governments can accumulate a significant portfolio and use the future earnings to grow and support a basic dividend fund or make a lump-sum payment to every young citizen. And of course, we should reconsider what we tax in the first place—but always ensuring that the rich pay their fair share when considering both income and wealth. In some cases, income tax can create additional barriers to creating jobs, making people more expensive to hire than robots. We might instead rebalance priorities to also tax things that negatively affect employment, for example, related to new technologies.

Overcoming Barriers to the Equality Levers

One significant barrier toward greater equality is that any major change in inequality will require those with power and those with the most access to power to support the change. On the face of it, this may seem an insurmountable challenge—like a turkey voting for Thanksgiving. Wealthy donors swell the coffers of political parties, ensuring they have a stranglehold on politicians. This political interference needs to be addressed, and a fairer playing field created.

Meanwhile, there are some signs that the tide is changing. The leading mouthpieces of the business elite and capitalism more broadly—the *Economist*, the *Financial Times*, and the World Economic Forum—have consistently shouted to anyone who will listen that: inequality is deeply destabilizing and must be reined in, climate and other environmental challenges are systemic challenges linked to issues like inequality that require new economic solutions, and that the business and finance community support stronger action from governments. And groups of super-rich, like the United States' Patriotic Millionaires, are vocally calling on governments to tax them more. While these groups remain in the minority, they do show that there is greater acknowledgment among the rich of the benefits to all of greater equality.

Another perceived barrier is affordability. How can countries afford the increased financial burden on governments? Well, much of what we have discussed is not a question of affordability; it is a question of distribution and allocation. In the long term, this will be achieved through fairer, higher taxes on the rich. In the shorter term, to protect the poorest in societies governments with their own stable sovereign currencies can spend money into existence. This economic strategy was successfully used to shore up the economy during the global financial crisis and again during the pandemic. There is no reason it cannot be employed a third time to create essential economic security during a deep transformation.

A final barrier is the prevailing narrative, which perpetuates the myth that inequality is a necessary consequence of creating a "better" world. We must just live with it, the story goes; it's the "natural" order in a capitalist society.

We need a new narrative that emphasizes the reality: Extreme levels of inequality are deeply destructive, even to the wealthy. They hold societies back. They create division and resentment. They breed conditions that are dangerous to everyone. They undermine democracies.

Conversely, democracies are stronger in more equal societies. The wellbeing and health of people is higher in more equal societies, as can be seen in the data for parts of Europe, Japan, and elsewhere. More importantly, we believe that the best way to uphold democratic values and deliver food, energy, and economic security for all is a profound redistribution of income and wealth.

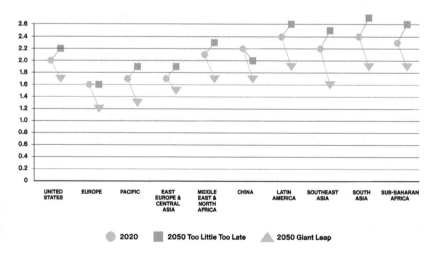

Figure 4.5. The Inequality Index for ten world regions in 2020 (*circle*), 2050 in Too Little Too Late (*square*), and 2050 in Giant Leap (*triangle*). The Inequality Index vertical axis is relative to 1980 = 1. Source: E4A-regional-220401.

Conclusions

These are big ideas for the Anthropocene. We think these are the ideas that span the chasm between today and the Earth for All economy. These are the ideas that provide the safety net of economic security across the gap during the transformation period. We have to acknowledge the next decade will be disruptive. Without a safety net people will dig in their heels, voters will be more likely to turn toward populist leaders, and citizens will reject a transformation that may feel like another attempt to line the pockets of the elite. But these safety nets can also be seen as innovation nets, allowing people a little more flexibility to create the future economy.

In summary, governments should pull hard on these levers: more progressive taxation, trade re-unionization for workers' negotiating strength, and a Citizens Fund with a fee and divided system. These measures can contribute to reversing the historic decline in worker share of output, as figure 4.5 demonstrates.

The solutions above address how to shift from the current economic paradigm, characterized by growing long-term inequality and divisions within society, to a new paradigm with greater equality and more trust within societies and therefore the possibility of more effective governance. Very simply, they are enablers, a foundation and a catalyst for all other turnarounds. More equal countries are more likely to support greater overseas development. They are more likely to support gender empowerment and investment in health and education. Or food and energy transformations that enable a regeneration of nature's ecosystems. This is because they are more likely to support an active, confident government to take real long-term decisions.

The Empowerment Turnaround

"Achieving Gender Equity"

This turnaround is about gender equity, women's agency, and championing families in a changing world. What do we mean by this? Improving women's access to education, economic opportunities, and dignified jobs, and all life's chances that these bring, will build better, stronger, resilient societies. It will also determine the future trajectory for humanity and our planet this century.

Discrimination against women's rights to equal education, equal pay, and financial security in old age is still pervasive around the world. In the empowerment turnaround, women gain better access to:

- Education, health services, and lifelong learning.
- Financial independence and leadership positions.
- Economic security through a universal basic dividend, or similar, and expanded pension schemes.

Together, this will accelerate the shift from discrimination toward greater gender equity and women's agency in society, a necessary step toward truly valuing our collective future.

More broadly, championing families means valuing whatever family or household structure people want. Families, large or small, can include childless couples, LGBTQ+ parents, multigenerational households, and all the diverse forms of human bonds in everyday living arrangements. Families need secure incomes, access to universal healthcare, flexible working, adequate pensions for all, and humane parental leave. These are the essential foundations for a functioning economy that drives human progress.

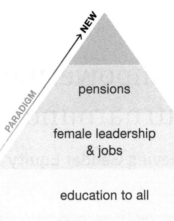

Figure 5.1. The empowerment turnaround brings the many benefits of gender equity to societies. A Giant Leap means to immediately create more opportunities and equality for women and girls, from school via work life to old age.

Supporting the empowerment of women can easily undermine itself if it leaves others behind or fails to understand the intersectionality of discrimination. A more empowered society means the specific context and needs of all marginalized groups, such as Indigenous groups and refugees, are understood and policy interventions address them. This includes an understanding that men are also discriminated against for reasons of race, sexual orientation, religion, income, and so on and cannot be left out of gender empowerment policies. Earth for All Transformational Economics Commissioner Jane Kabubo-Mariara, president of the African Society for Ecological Economists, also notes that empowerment of girls at the expense of boys will backfire: "In Kenya for instance, there is a general feeling that since the Beijing Women's Conference of 1995, too much affirmative action has given the girl child an upper hand over the boy child, and there are calls to reverse this." Boys are not to be left out. In a culture of toxic masculinity, leaving men out without addressing their anxieties has led to catastrophic gender-based violence and femicide, notes Club of Rome co-president Dr. Mamphela Ramphele. Hence this turnaround is toward removing discrimination in order to rapidly transition toward inclusiveness and gender equity.

Public investment in universal education during a phase of tumultuous change must therefore be seen as a top priority. But not just any "education." This turnaround also includes rethinking educational systems to bring them out of a worldview steeped in thinking from the Industrial Revolution (and largely designed to suit boys) and into a worldview that values lifelong learning and connectivity between people and ecosystems. This means equipping girls and boys with the cognitive tools they will need to navigate this century: critical thinking, systems thinking, and adaptive leadership in order that they can thrive in a world undergoing deep transformation.

Public investment in healthcare for everyone has been shown time and time again not only to provide the economically optimal long-term healthcare and wellbeing for the most people but also, as seen during the pandemic, to help build trust in the role of governments to protect societies. As economist Mariana Mazzucato points out, "In 2020, global GDP grew by $2.2 trillion as a result of governments increasing their military spending; meanwhile, the world still has not provided the mere $50 billion needed to vaccinate the global population."[1] Ultimately this turnaround requires states to become more active in the economy to help build this future. A simple starting point is to set a goal of universal education and healthcare and work backward from there to work out how to achieve it.

Equal status in the workforce is an essential goal. Women are roughly half of the world's population yet remain on the losing side when it comes to income and wealth. In a gender-equal world, women would of course earn roughly 50% of all labor income. But overall, women's share of total incomes from work (labor income) was 30% in 1990 and still stands at less than 35% in 2022.[2] And less than 20% of landowners in the world are women. It's not just that women tend to earn less than men. Women are disproportionately stuck in low-paying jobs and face glass ceilings that keep top jobs just out of reach, perpetuating the problem. And, last but not least, women are still too poorly represented in politics, finance, boardrooms, and executive teams. Gender equity is, therefore, essential for resilient, healthy societies. If achieved, it brings important secondary benefits to societies...and the planet.

Population

Possibly the easiest way to start a long, heated debate is to mention global population growth. Thomas Malthus famously ignited furious arguments 220 years ago. These arguments were still raging in the 1960s when Paul and Anne Ehrlich added fuel to the fire with their bestseller *The Population Bomb*. They were onto something. In just under fifty years, the global population had almost doubled from 2 billion people to reach 4 billion in around 1975. In 2022, the human population has almost doubled again (7.9 billion people) and growing at around 80 million per year.

So, when will the population double again to 16 billion?

It will not. It will not even get close to this number. In short, the good news is that the "population bomb" that many feared has been defused. The last forty years has seen a huge change in demographics. The growth rate peaked in the sixties and has been falling steadily since. Around the world, women are having fewer children. In fact, the average number of children per woman in 2020 was just above two. This hides a large divergence across the world. In places like Japan and South Korea, it is less than two. In low-income countries and in particular in fragile states, the numbers are much higher.

Despite progress on slowing population growth, if the world continues on current trends, the UN's median population projection sees a peak at around 11 billion people by the end of the century.[3] The additional pressure this would put on the Earth system is significant, and it could make or break societies. The UN forecasts this growth to mainly happen in Africa. In some parts of West Africa, births per woman are still as high as six or seven children. Currently, 1.3 billion people live on the continent, and this could double according to the UN median forecast, which we believe is too high. Our model supports our view that the world population will peak around some 9 billion people around 2050.

This, in turn, means we can avoid a perfect storm. Birth rates are connected to many factors including better education, better health, more jobs for girls and women, higher income per person overall, and accessible contraception. All of these lead to more freedom to choose how many children women desire.[4] Yet recent data estimates that 222 million women in low-income countries had an unmet need for family planning. Implementing these solutions in sub-Saharan Africa and other low-income countries would avert more than 1 million infant deaths and 54 million unwanted pregnancies that, if not prevented, would result in 21 million inadvertent births, 7 million miscarriages, and 26 million abortions of which 15 million would be unsafe.[5] The Giant Leap aims to provide these solutions to help address some of the health and social challenges related to poor family planning provision.

Our analysis indicates that the main actions that make up the empowerment turnaround (education, health, income, pensions) will lead to smaller families, longer lives, and a global population peaking possibly below nine billion people by around 2050. Then population will start a slow and steady decline during the whole second half of the century. How can this more steady situation come about?

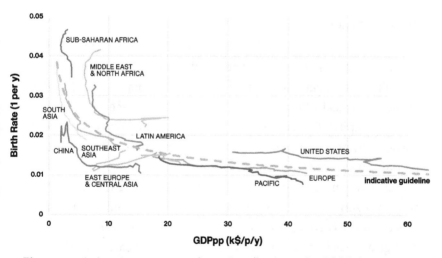

Figure 5.2. As incomes per person have risen (horizontal axis), birth rates have come down quickly in all regions. Lines show historic 1980–2020 data. Dotted line shows the indicative guideline for future birth rates toward 2100 as a function of GDPpp. Sources: Earth4All analysis, based on data from Penn World Tables; UN Population Division.

Number of Children per Woman

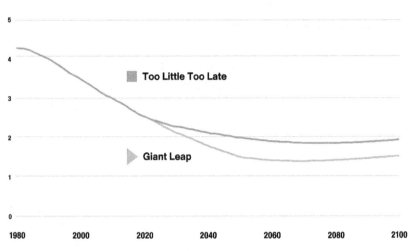

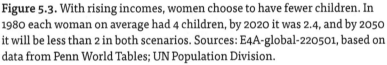

Figure 5.3. With rising incomes, women choose to have fewer children. In 1980 each woman on average had 4 children, by 2020 it was 2.4, and by 2050 it will be less than 2 in both scenarios. Sources: E4A-global-220501, based on data from Penn World Tables; UN Population Division.

Turning It All Around

On first glance, the key challenges grouped in this turnaround may seem like a ragbag of unconnected problems: rethinking education, providing healthcare for all, and reversing overpopulation. But they are all intrinsically linked to one central idea: gender equity. It is easy to show empirically that societies that invest in greater gender equity, particularly promoting equality of economic opportunities and social mobility, have better outcomes for all.

As this century progresses, societies will increasingly have to adapt to aging populations. This will require profound restructuring of economies. On top of this, societies need to adapt to environmental change and social transformations. Universal education and health-care are foundational investments for resilience and to instill trust in governments to bring all citizens on this journey. Put simply, Earth for All requires more active governments, and this will not happen unless the majority feel they are benefiting from the arrangement. The good news is that we are not starting from scratch.

Of all the turnarounds we will discuss, gender equity and agency has made the most progress in the last fifty years. Admittedly, the world started from a very low baseline, and there are mountains yet to climb. Nevertheless, five decades of progress has actually seen the education gap and even the pay gap between men and women shrink in most places. And it is becoming more common for parental inheritance to be gender neutral.

But the pace of change is clearly not fast enough to deliver the Giant Leap scenario. Most significantly, without a renewed push, populations may grow to levels that will be profoundly challenging for all of humanity.

All five extraordinary turnarounds will in some way improve gender equity and agency, bringing a multitude of additional benefits. For example, availability of food and energy brings greater economic security, influencing long-term decisions relating to families. But the biggest factor when it comes to making different choices is not rocket science: it's financial independence. Financial independence gives the freedom and the power to be able to say no to men, no to demeaning labor, no to unwanted marriage. And say yes to education, training, career, and control over fertility.

Looking around the world, gender equity and valuing family life is at least part of a recipe for economic success, it seems. The wealthy Nordic countries of Denmark, Finland, Iceland, Norway, and Sweden regularly top international polls on wellbeing and happiness. These are market economies with highly effective states that are committed to investing in families. The Nordic countries dominate the World Economic Forum's Global Social Mobility Index (2020); a child born in Denmark has a better chance of living the American Dream than a child born in the United States. Of course, these countries are far from perfect, as they have very high consumption footprints. But interestingly, citizens have a high degree of trust in their governments, and this has allowed these countries to make effective long-term decisions that benefit all; for example, these countries were among the first to commit to net-zero carbon emissions. More recently, other countries have more actively embraced the notion of wellbeing economies. An alliance has formed that includes New Zealand, Scotland, and Wales, as well as Finland and Iceland, to promote new economic ideas that work for the majority. At the time of writing, these countries fiercely advocated for gender balance in government, and all are led by women.

The wellbeing economies are now grabbing the attention of other countries. The governments of Costa Rica, Canada, and Rwanda are exploring these approaches to economic prioritization. And in other developments, some cities have begun to explore economic models like the "doughnut" championed by economist Kate Raworth. They are striving to operate their economies within both biogeophysical planetary boundaries and social boundaries relating to inequality, health, education, gender, and more.

There is no perfect economic system. But there is now a vibrant ecosystem of powerful, transformational economic ideas that empirically work in practice. Common to all is commitment and investment in gender equity and agency.

Transforming Education

One of the most important policy interventions to address population growth is investment in education. Education is the best escape route from a life in chains. It provides social mobility, economic security and opens up a world of opportunity. Educating girls increases their lifelong earnings and national income, reduces child mortality and maternal mortality, and helps prevent child marriage. Progress in the last fifty years means that in many places around the world there is close to gender equity. In some, there are now more girls than boys in education. However, Africa and South Asia have only just reached the level that current high-income economies reached in the early part of last century.

Education is often measured simply by the number of years at school. This works up to a point. Just as higher incomes are often conflated with better lives, despite evidence that life satisfaction stops growing above a certain income level, we find education is often conflated with schooling. But schools have changed surprisingly little since the Industrial Revolution. If you google "classroom," though the technology has changed over two centuries, the basic idea is intact; you see a universal archetype of rows of desks facing a teacher.

Our Schools Were Designed for a Different Era

It would seem to be a strange form of empowerment for women and girls to be inducted into a schooling system that would not look out of place in patriarchal nineteenth-century France. In the most common model for schooling, children are still repeatedly contained for standard defined periods of perhaps an hour and given things to be memorized, often without an obvious reason to do so. Examinations usually focus on remembering these things. Much of what is learned is more of a rite of passage and rapidly forgotten. Beyond literacy and numeracy, the real learning is in the social interactions between pupils and learning about the tricky business of growing up. But of course, in many rural schools, just being there, having a teacher who is competent and toilets that are functional, can seem like a triumph of the everyday.

Beyond this lies the dream. The structured ladder of schooling, often into higher education, offers something genuinely priceless: social mobility. And this is why it is valued, for with social mobility comes hopes for a better life not just for the learner but for the whole family. But, just because the ladder is climbed, and there are now whiteboards not chalkboards and computers along the back of many classrooms, this does not signal better *educational* outcomes. If these five turnarounds have illustrated anything, it is that a mindset based on nineteenth-century reductionist and linear causal relationships, as if the best way to build knowledge is to assume the world is like a machine that can be understood from the parts, is a big part of the problem.

The overhaul of education everywhere should build on two foundations: critical thinking and complex systems thinking. Arguably the biggest challenge in the world today is not climate change, biodiversity loss, or even a pandemic. It is our collective inability to tell fact from fiction. In democratic societies, misinformation and disinformation had been kept at bay, to some extent at least, by checks and balances within mass media. Social media smashed this model apart. It has industrialized the spread of misinformation and disinformation in the world, polarizing societies, reducing trust, and contributing to our shocking inability to cooperate around common challenges, or even agree on the interpretation of basic facts. During the pandemic, mask wearing, or not, in some countries became a politically polarized issue where empirical evidence was ignored and ridiculed. This contributed to many more unnecessary deaths. Failures like this arise from a systemic issue that will require long-term solutions. Education systems have a duty to step up and teach critical thinking to help the next generation navigate this information minefield.

The second foundation for education is complex systems thinking. The Earth4All model is built on system dynamics and systems thinking, tools *The Limits to Growth* pioneered fifty years ago. Most real-world systems are complex dynamic systems, whether ocean and climate or urbanization and stock markets. So, an education system that largely ignores these bedrock features until university is obsolete. Knowledge systems used by many Indigenous peoples often embrace a systems view, a complexity view, and narrative approaches to learning. These approaches could be incorporated into new curricula based on systems thinking. Both of these foundations are critical to one of the fundamental skills necessary for navigating the future: adaptive leadership, or the ability to take decisive, informed action in rapidly changing circumstances.

Cost Still Bars Millions of Children from Education

The question of education is a systemic economic challenge. In the 1980s, a debt crisis struck many countries in Africa. The International Monetary Fund and the World Bank stepped in to lend money to cash-starved states. But this money came with conditions. They insisted that countries control public expenditure. These international demands translated on the ground to schools introducing user fees. Paying for access to schooling became widespread. A UNICEF study from the time found that the poorest 40% of families spent over 10% of their yearly income just to send two children to primary school in about half of low-income countries assessed. Recent UNESCO figures put educational exclusion for all reasons around the world at 258 million children.[6] The global pandemic has certainly increased this number, but at the time of writing (2022) no one knows by how much. Closures in the first two years of the pandemic lasted roughly twice as long in low-income countries compared with high-income economies. And the adverse impact of this shock is magnified because the share of those at school age in low-income countries is nearly double that of high-income economies.[7]

The good news is that there are many programs offering alternative models of schooling, as Dr. Mamphela Ramphele points out.[8] They are adapted to different cultural and geographic settings and more suitable to the needs of today and the future, not the needs of yesterday. One example is the LEAP schools in South Africa, which ranks as one of the most economically unequal countries on Earth. The LEAP schools are designed to address inequality by providing free education to the most marginalized communities. The curriculum is designed to inspire and engage and instills a spirit of personal agency and connected global citizenship. Around 80% of LEAP students have degrees or diplomas or are continuing with higher education. They say, "We are proving the impossible: that no matter how deprived, any child in South Africa can graduate from high school, obtain a tertiary qualification, and look forward to a fulfilling future." Their success has led to the creation of a new LEAP Institute to facilitate diffusion of the model across Southern Africa, notes Ramphele.

Financial Independence and Leadership

There are many ways to create economic freedom and security beyond employment. What if all women in a community received unconditional monthly cash payments? In trials in India, women received a form of a universal basic income. The goal was to explore its impact on poverty and to support greater empowerment and agency. Assessments of the trials concluded that the supplemental income delivered three benefits: Nutrition improved in the women's families, meaning that health also improved and children spent more time in school; The trials had a positive impact on economic growth; and, the income gave women greater control of household spending decisions.[9] It is important to emphasize that this approach did not involve someone else deciding what a woman needs—as in so many paternalistic welfare programs. Providing a stable source of income, which is not means-tested or conditional, fosters agency, equity, and inclusiveness.

The poverty turnaround chapter gave insights about creating self-sustaining prosperity within poorer nations. Done well, this could mean more government income could be available for better schooling. Education could be made free, accessible, and universal. However, when it comes to gender, the challenges are more than getting and keeping a place in school. Cultural expectations and obligations weigh heavily on girls and women in many parts of the world. Many are—despite education—denied access to jobs.

Alongside universal basic income and free education, we strongly support universal health coverage as a systemic solution toward greater gender equity. In the twenty-first century, we consider this an essential human right and foundation for a functional society. In places like the UK and Sweden, an effective, free healthcare system instills trust in governments that the wealth of the nation is shared more fairly among all citizens. This is true commonwealth.

Because universal health coverage is based on a systems approach to healthcare provision across all of society, it can invest greater time and resources in preventative interventions. Investment in prevention is usually a small proportion of healthcare expenditure, notes Earth for All Transformational Economics Commissioner Andrew Haines.[10] Examples include education about diets and physical exercise as well as structural changes in society to make healthy choices more feasible and accessible. These actions can lower the overall costs of healthcare, and help people make choices that benefit their long-term health. It also provides additional economic security for those most vulnerable in society.

If the generally positive trends outlined in this chapter accelerate beyond the historic trends, then we should expect more gender balance in leadership and power in business and in government. More actions and regulations are needed to increase the diversity of people in larger roles in economic and political life too.

A Secure Pension and Dignified Aging

One of the main reasons the global population will keep rising for some more decades in both our scenarios is because many of us are still young (global median age was around thirty in 2020), and many more people live to a ripe old age. An aging population requires greater spending on health and long-term care, it shifts the burden of disease, and it may lead to shortages in the labor force unless the pension age is raised in tandem with rising life expectancies. Gaps in pensions, where they exist, cause income insecurity. An aging population requires greater investment in welfare provision and this puts pressure on the workers in the economy. But the accompanying reduction in the number of young will reduce the pressure.

A starting point to address the challenge of an aging population is to simply increase the retirement age in line with increases in longevity. This can help reduce the financial burden on the national workforce. Of course, this brings with it additional challenges: retirement has always been a natural way for others to advance their careers. Beyond this, financial security in old age is essential, which is why we support an expansion of pension provision especially for women.

The day for big ideas like an expanded pension, a universal basic income, and a universal basic dividend is finally here. They could have a profound impact on wellbeing, gender agency and empowerment in

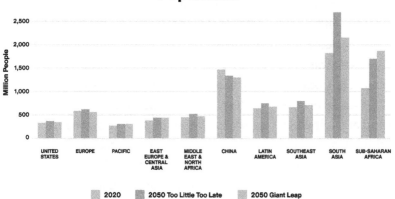

Figure 5.4. Population per region in 2020 and in 2050 for Too Little Too Late and Giant Leap. Sources: E4A-regional-220427; Penn World Tables; UN Population Division.

all countries. This is the moment to take bold decisions to implement them. They will not only help redistribute income and wealth more fairly, they will provide essential economic protections through what we know will be an economically turbulent time of transformation. These ideas are linked to the extraordinary turnarounds of inequality and energy, and are further discussed in those chapters (4 and 7) as well as in the new economic thinking chapter (8).

Conclusions

The starting point to valuing our future is valuing equality, diversity, and inclusion. The empirical data shows that economies that support greater equality score highest in international league tables of wellbeing and human development. These are conditions that also enhance economic competitiveness. But more importantly, these are the conditions that enhance resilience to shocks like financial crises, pandemics, and food price volatility. They help build social cohesion because fairness and justice are valued. We will need wartime levels of social cohesion in the coming decades to deliver the future we want.

A significant barrier to progress is, of course, culture. Patriarchal societies have dominated for so long that every single aspect of many societies' art, music, commerce, and politics is warped by a mindset steeped in male hierarchy. It creates a powerful narrative based on a single idea: "This is the natural order of people." Bit by bit, the patriarchal hierarchy is being broken down. This will take generations to disappear entirely.

Gender equity brings a profound additional benefit. In the last fifty years, the once exponential curve of population growth that dominated from 1800 to 1975 has bent down. This is an astounding achievement of economic development. But huge challenges remain ahead of us, the most significant being to provide a good life for all within planetary boundaries. This can best be achieved by ensuring that the world population stabilizes at about 9 billion around 2050, and is allowed to decline toward 2100 because families choose to have fewer children when they see a prosperous future. This is the ambition of the Giant Leap scenario.

6

The Food Turnaround

Making the Food System Healthy
for People and Planet

The last fifty years has witnessed an astonishing turnaround in food security. Starting in the 1970s, the world has seen a dramatic fall in the number of deaths from famine even as the global population doubled. Of course, far too many people have suffered and died, and far too many still lack food security, but it is still important to recognize steady progress.

However, progress has come at a cost. The way we farm, transport, and consume food affects more planetary boundaries than anything else. The agriculture sector is one of the biggest sources of greenhouse gas emissions. It is the largest driver of deforestation and biodiversity loss, by far the world's largest sector consuming freshwater, and excess fertilizers leak into air and streams, lakes, and oceans—causing vast dead zones and even more global warming.

So, agriculture is certainly not working for the planet, that is clear. But it is not working for people either. We have increasingly moved further away from local production for local consumption and become alarmingly dependent on a few major food-producing countries.

Nearly one in ten people (9%) worldwide remain severely food insecure with 821 million people undernourished. On the flipside, an astonishing two billion people, one quarter of the planet, are now overweight or obese.[1] In 2017, 8% of deaths worldwide were attributed to obesity.[2]

The extraordinary food turnaround focuses on three groups of solutions.

change diets

food-system
efficiency

new farming techniques

NEW
PARADIGM
CURRENT

FOOD

Figure 6.1. The food turnaround: Regenerative agriculture and sustainable intensification start creating healthier soils and ecosystems, while consumers shift from grain-fed red meat to nourishing, healthier diets and the industry tackles wastefulness across the entire food system.

The way we farm food and other commodities needs rapid and extensive reform. Regenerative agriculture and sustainable intensification can make farming more efficient, increase yield, and reduce inputs of chemicals that are harmful. These are different approaches, but what is called "sustainable intensification" can be a useful bridge to more regenerative practices. Farms can do far more with much less. Land expansion must stop and degraded land regenerated to protect priceless biodiversity and carbon sinks. Farms themselves must become vast stores of carbon, not vast emitters. And out on the high seas, fish stock collapse must be avoided, while coastal aquaculture projects must contain their pollution and their encroachment on marine habitats.

The well-fed must adopt healthy, lower-impact diets, while the malnourished and undernourished must be lifted out of their predicament with regeneratively grown, healthier foods. People everywhere need access to safe, nutritious food that is produced within the planetary boundaries.

We must tackle food waste along the entire food chain from production, distribution, and shops to consumers' tables and bins. About one third of all food is wasted between the field or fishing net and the fork. Eliminating just 25% of that would free up enough food to feed all people on Earth.

This essential transformation to a new food system paradigm will be one of the most dramatic changes in the history of our species.

Consuming Earth's Biosphere

In the future, as population grows, we need to tread very, very carefully. Humanity is slicing, dicing, and simplifying Earth's biosphere—our life support system—in the service of human food production and material consumption. By mass, about 96% of all mammals on Earth are either humans (36%) or our livestock (60%), mainly cattle and pigs. Just 4% of mammals are now wild.[3] Or put another way, our livestock outweigh wild mammals 15 to 1.

The scale of appropriation and consumption requires vast tracts of land. On Earth, glaciers and ice sheets make up about 10% of the land surface. About 19% is taken up by barren lands—exposed rock, desert, and salt flats. The remaining land (71%) is described as habitable. Humanity has commandeered about half of this habitable land for farming, and we have modified or interfered with much of the rest in some way.[4] For livestock production alone, we use an area of land equivalent to North and South America combined. In the ocean, around 90% of fish stocks are either overexploited or fully exploited.[5] A rapidly growing aquaculture sector is taking up more coastal space each year. When it comes to the air, about 70% of birds by mass are farmed poultry, and the rest wild. We live on a planet of chickens.[6]

But the impact of food production and deforestation goes beyond the swallowing up of life on Earth. About one quarter of all greenhouse gas emissions comes from land use. Agriculture is responsible for about 70% of all water withdrawals. The final major impact is pollution: the cause of growing aquatic dead zones is our overuse of fertilizers. Agriculture is linked directly to 78% of eutrophication in lakes, rivers, seas, and the ocean.

It turns out the scale of impact is neither necessary nor does it particularly serve up healthy diets.

Today's food crisis is only half the story. Tomorrow we face even more challenges. Given the current food system, the UN Food and Agriculture Organization estimates that the world needs to produce about 50% more food by 2050 to feed a population that is growing in size and wealth.[7] While some dispute that additional food production is this high, demand will undoubtedly increase in the next three decades while the food system will increasingly be hit by more and more extremes. It is virtually certain that wet areas on Earth will get wetter and dry areas drier this century. This means more floods in regions prone to flooding and more droughts in drought-prone regions. Extreme heat will increasingly damage plants. All of this will be increasingly disastrous for food production. Historically, civilizations thrive or fade, live or die, depending upon how their leaders manage scarce water resources for food production.

The food system is not just unsustainable, it is also highly fragile, dependent on tight global trade in monoculture crops, fertilizers, and fossil energy. Staple food crops—like grain, meat, and oil—are traded globally, with many countries heavily dependent on imports from a few breadbaskets like Russia, Ukraine, Australia, Argentina, and the United States. Phosphorus fertilizers often come from Morocco (West Sahara), the United States, and China. Nitrogen fertilizers often come from countries rich in natural gas, like Russia and Ukraine. Oil to run farm equipment is sourced from a few nations. This creates supply chain bottlenecks if these countries are hit by disruptions such as harvest failures or war.

This broken system impacts food prices, too. In 2022, global food prices reached their highest ever level following the Russian invasion of Ukraine. Cereal prices alone rose 17% in one month. Global crop and economic models project that cereal price could increase by as much as 29% by 2050 as a result of climate change unless there are immediate and deep cuts in emissions. But more than that, extreme events like prolonged droughts have cascading impacts.

Societal unrest is tightly connected to the price of bread and other foods.[8] During the Arab Spring of 2010 and 2011, high food prices were an important factor in driving people onto the streets to protest and eventually topple governments across the region. At that time, droughts in Russia, Ukraine, China, and Argentina severely cut wheat harvests while torrential rains in Canada, Brazil, and Australia had a similar effect. Grain prices skyrocketed.

The link between societal tension and food prices is particularly acute in low-income countries. When prices of many commodities, like oil, go up, people use less, but food consumption is "income inelastic." Or, to cut the jargon, people have to eat regardless of how much money they earn. According to recent research, in low-income countries, increases in international food prices lead to a significant deterioration of democratic institutions and a significant increase in the incidence of anti-government demonstrations, riots, and civil conflict.[9] The slogan on the streets of Cairo during the Arab Spring uprising in 2011 was "Bread, Freedom, Dignity," in that order. Of course, another risk—and legitimate coping strategy in times of rising social tensions, poor economic opportunities, and conflict—is migration. When this happens, migration can have spillover effects contributing to rising social tensions and political unrest elsewhere.

The multiple breadbasket failures seen in 2010 were a shock. Was this event one of a kind or can we expect more as Earth heats up? Science is getting a better handle on the risks related to breadbasket failures and what this might mean for the future. The jet stream is a band of fast-moving air that swirls around the northern hemisphere above the most important grain producing areas of Asia, North America, and Europe. It is slowing down and beginning to act weirdly as the planet warms up. Weather systems that are pulled along beneath the jet stream can stall, intensifying weather conditions. Where once a high-pressure weather system might sweep into Europe bringing warm weather for a few days, now these high-pressure systems sometimes stick around for weeks bringing devastating heatwaves. But more than that, these weather systems—bringing rain or drought—can stall in several regions simultaneously. The failure of multiple breadbaskets across the world is now one of the biggest risks facing food production and keeps climate scientists awake at night.

We face a triple challenge in agriculture: to produce more healthy food, without destroying the planet, while building resilient production systems that are able to withstand rising shocks. As we look out over the horizon, we know demand for food is driven by both population growth and income growth. We know that we are increasingly dependent on only a few food-producing countries. We know income also drives dietary preference toward the all-consuming Western diet. We know agriculture is limited by the inevitable scarcity of land, water access, and poor soil quality. And we know climate change will affect yields and the spread of diseases affecting crops and livestock. Throw in several black swans long the way—high-impact shocks that are difficult to predict—and this reinforces the essential need for a food-system turnaround to manage for fragility, volatility, and risk and to rebuild for resilience, price stability, and wellbeing. While there are no simple solutions, we believe the following three proposals are the most significant levers to drive large-scale change. We are not naive enough to think these are the only levers, but when it comes to human wellbeing, respect for planetary boundaries, and reducing social tension, these give the most bang for the buck.

Solution 1: Revolutionize the Way We Farm

The legendary biologist E.O. Wilson proposed using no more than half of Earth for human needs saying, "The Half-Earth proposal offers a first, emergency solution commensurate with the magnitude of the problem: By setting aside half the planet in reserve, we can save the living part of the environment and achieve the stabilization required for our own survival." The Half Earth proposal is supported by the science of planetary boundaries. Having transformed and exploited approximately 50% of natural ecosystems on land, we have transgressed the boundaries on land system change (forests), chemical pollution, biodiversity loss, and fertilizers.

The only way back into a safe operating space is to stop agricultural expansion into remaining forests and wetlands and to use water and nutrients more efficiently. We have arrived at the red light. This has major implications for how societies farm in the future, how we make our food system resilient, and ultimately how we feed more people without growing our environmental footprint.

We propose six principles for a new era in food production in the Anthropocene. First and foremost, no more expansion of agricultural lands into forests, wetlands, or other ecosystems. We must grow more food on less land and regenerate degraded land. Second, farms must become stores of carbon, not vast emitters—within the next decade or so. Third, our farms must enhance the rich diversity of life. Fourth, the future of our civilization depends on the health of our soil. We must restore our soils to good health. Fifth, we must manage our ocean and freshwater resources for resilience. And sixth, we must support more local production for local consumption where possible.

Meeting these principles means revolutionizing the failing food system. The modern farming system is really a high "through-flow" system. Fertilizers, other chemicals, and water flow through the system driven by energy from fossil fuels and pumping waste onto land and into waterways and the atmosphere. The system needs to move from linear to circular, from destructive to regenerative. The good news is that many of the solutions to feed a population of 9 billion people exist already and are found within farming approaches gaining popularity in recent years. But we must admit, of all the turnarounds, the future of the food system has generated the strongest debates. How much emphasis should be placed on organic farming or lab-based alternatives to beef? And, how do we minimize the harmful impacts of artificial fertilizers and other manmade chemicals but also reap their benefits? This is one reason we suggest creating a Food System Stability Board at the end of this chapter.

Key among those approaches are regenerative agriculture and sustainable intensification. Both are lighter on the planet than conventional agriculture. Sustainable intensification focuses on maximizing crop production while improving ecosystem protection, employing modern technologies, and adopting circular waste-reducing practices. "Regenerative agriculture" is an umbrella term for a range of agricultural systems that place emphasis on building soil health, carbon stores, and crop diversity while restoring ecosystems. Hunter Lovins, president of Natural Capital Solutions, says regenerative agriculture delivers increased soil health, increased farmer and community health, greater economic resilience, better water conservation, dramatically more carbon in the soil, and increased biodiversity.[10]

Regenerative farmers use a range of techniques, including cover crops, crop rotation, and composting to build healthy living soil. To protect the valuable carbon in soils, and to look after fungal networks, microorganisms, worms, and other soil builders that lie beneath the surface, they do little or no tilling. Rather than plowing fields, they drill seeds into the soil.

More farmers are adopting these ideas. After several years of freak storms and crop failures, North Dakota farmer Gabe Brown saved his farm using regenerative agriculture. He raised the concentration of carbon in his soil from a little under 2% organic matter per acre to over 11%.[11] Over two decades, he helped rein in climate change and increase yield at the same time.

Regenerative farmers also use grazing practices that regenerate ecosystems. They do this by creating an environment where livestock mimic the role of wild ungulates on the landscape. Some farmers plant trees on agricultural lands, which brings many benefits: it reduces erosion, it helps recharge groundwater, and provides shade for livestock while also producing fruit, nuts, or timber.

These techniques improve soil health, and healthy soil retains large stores of carbon.[12] Regenerative agriculture done right can also increase biodiversity, cycle nutrients through the ecosystem, filter water, and provide other environmental benefits. And grass-fed livestock offers an alternative to the factory-farmed, grain-fed livestock operations whose effects on the environment, health, and animal-welfare are well-known.

Regenerative farming also increases food resilience. By increasing soil fertility, using local seed varieties adapted to local conditions, farmers can achieve greater yields with fewer inputs, and they become less susceptible to crop failures as climatic variations intensify.

In 2015, Vijay Kumar, a retired civil servant from the agricultural department, began working with smallholder farmers in India to explore how they could become more resilient and more profitable. He now works with millions of farmers in nine Indian states to help them move to regenerative practices, what he calls Indian Community Managed Natural Farming. He is helping smallholders adopt principles such as: minimal disturbance of the soil, biomass covering the soil, and living root always in the soil. One big win for Community Managed Natural Farming is increased soil moisture. Farmers can harvest crops year-round, potentially tripling their income, while sequestering large amounts of carbon.[13]

In Africa, Million Belay[14] calls a similar approach agroecology. Across the continent, Belay believes it can help poor farmers double their productivity, bring real food security, and help store carbon to areas previously vulnerable to famine.

Achieving the food turnaround at the requisite pace and scale will require a mix of agricultural approaches and the adoption of sensible technologies. Sustainable intensification allows agricultural yields to increase while minimizing adverse environmental impact and without the conversion of additional nonagricultural land. Sustainable intensification does not promote any particular method of agricultural production.[15] The aims can be achieved using pesticides or not, or using artificial fertilizers or not.[16] On this last point, one thing is clear: in high-income countries overuse of artificial fertilizers is destroying ecosystems and must be reduced. However, in low-income countries, the problem is low yields, soil health, and a lack of adequate fertilizers. More fertilizer is needed desperately, at least until there is food security and the soils are fully regenerated, but new approaches to farming can substantially reduce fertilizer input. So reducing fertilizer use in high-income nations but increasing it in lower-income nations can increase food production without increasing the footprint. A true win-win. Sustainable intensification also prioritizes climate-resilient techniques, ever important in a future marked by drought and deluge.

Technologies from satellites and drones to moisture sensors and robots are revolutionizing farming. Fertilizers can be targeted with pinpoint accuracy using satellites to provide real-time data to farmers reducing runoff into streams and rivers. Irrigation can be more carefully monitored and managed in order to optimize the water used on the farm. One of the technologies with the quickest returns has been simply to add GPS to tractors so farmers always know where they have covered and where there are gaps. And in cities and towns, vertical farming can deliver higher productivity in smaller areas, with shorter growing times and less water. The priority, then, is to direct this technological revolution to support and promote healthy, sustainable diets with lower food footprints, and make this affordable for the millions of small farms worldwide.

In the Earth4All model, we assume these sustainable and regenerative approaches to farming are adopted more broadly decade by decade, leading to yield increases and other benefits related to soil health and biodiversity. This can be achieved through changes to agricultural policies, for example, subsidies to promote these farming approaches, education to drive behavioral change among farmers and the agricultural industry more generally, and exponential scaling of new technologies and knowledge exchange as the prices drop.

The Earth4All model allows you to turn a dial to increase the share of agricultural land that has shifted from conventional to regenerative practices with near-zero use of fossil fertilizers. Another dial influences the speed of the transition—how many years it takes to get there. To make Giant Leap, 80% of agricultural land must be transformed by 2050, up from a low baseline in 2020. In the Too Little Too Late scenario, just 10% shifts by 2100.

Solution 2: Change Our Diets

The Western diet—packed full of industrially grown processed meats, saturated fats, salt, corn-derived fructose syrup, refined grains, washed down with large volumes of alcohol—is taking over the world. Low on fruit and vegetables, this is both cheap and heavily marketed as aspirational, the diet of the middle classes, and viewed as a symbol of success and wealth. With its origins in the Industrial Revolution, the Western diet is associated with obesity, diabetes, cancer, and cardiovascular diseases.

The goal, then, is to shift away from this toxic diet toward richer, more varied diets that reduce risk of these diseases, and reduce the risk of destabilizing the planet. These changes must also be accompanied by fairer distribution that makes healthy food affordable and accessible to those living in urban food deserts and other underserved regions.

Our analyses show there is space on Earth, without expanding agricultural land any further, to feed at least nine billion people a healthy, nutritious diet. Indeed, our work builds on the seminal analysis conducted by the EAT-Lancet Commission on healthy diets from sustainable food systems. Ultimately, this means reining in overconsumption of industrially produced meat and dairy in some places, but it does not mean forcing a vegan or vegetarian diet on anyone.

A shift from the typical Western diet to a planetary health diet, notes Andrew Haines, means far more than just reduced industrially produced meat consumption. It involves a large *increase* in consumption of fruit, vegetables, legumes, nuts, and seeds. Many health benefits arise from these changes. With them, some ten million premature deaths could be prevented annually by 2040.[17] Much more importantly, poor nutrition still affects hundreds of millions of people, so a planetary health diet must focus on underconsumption as much as overconsumption.

The ocean is another important source of healthy food. Currently, 3 billion people get 20% of their animal protein from seafood. Sustainable aquatic food has the potential to prevent 166 million people from suffering micronutrient deficiencies. Global demand for seafood will roughly double by 2050, according to the 2021 Blue Food Assessment published in the journal *Nature*.[18] This demand will be met primarily through increased aquaculture production rather than by capture fisheries. But aquaculture must use sustainable practices.

Innovation will likely play an important role in the transition toward a planetary health diet. Plant-based alternatives to milk are growing in popularity and increasingly marketed and perceived as "aspirational" foods. So too are plant-based and "lab-grown" alternatives to beef and chicken. We are also heading for an innovation revolution in the field of "precision fermentation and cellular agri-

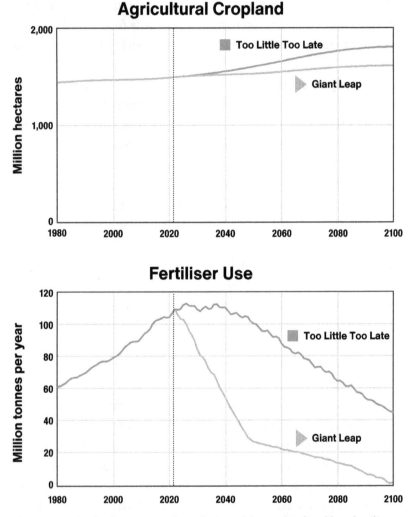

Figure 6.2. The food turnaround results in stable cropland and less fertilizer use, while feeding the world sufficient foods. In Giant Leap, agriculture no longer expands into natural areas, allowing forests to grow back. As sustainable intensification, switching diets to less grain-fed red meat and regenerative agriculture is scaled, the use of fossil fertilizer drops rapidly in Giant Leap.

culture"—in which microorganisms directly produce the kind of proteins we would typically get via a cow, fish, or bird. Precision fermentation uses the large and mostly untapped variety of yeasts, fungi, mycelium, and microalgae as hosts to produce ingredients identical to animal proteins, for example egg whites or dairy. The 2022 report from the Intergovernmental Panel on Climate Change concludes that emerging food technologies such as cellular fermentation, cultured meat, plant-based alternatives to animal-based food products, and controlled environment agriculture can bring substantial reductions in greenhouse gas emissions from food production. Of course, we must be careful not to replace one problem with another. As always, we need to look at the entire food system and adopt solutions that are holistic while ensuring that we feed as many people as possible nutritious food within planetary boundaries.

This diversity of new ideas and innovations is good as it pushes people to think more broadly about what they eat and why they eat it. Undoubtedly, the rise of these new industries in some places is a sign that a significant transformation is beginning. The Earth4All model allows us to test the effect of different assumptions concerning the future fraction of climate-neutral meats, whether grass-fed or "new meats." To achieve a Giant Leap, we must assume that 50% of all red meat is climate neutral by 2050. In Too Little Too Late, we see that only 10% has shifted by 2100, meaning that new meat remains a pretty marginal product in TLTL. The consequences for cropland and fertilizer use are shown in figure 6.2.

Solution 3: Eliminate Food Loss and Waste

With a growing population, potentially reaching nine or ten billion people, it makes sense to worry about our scale of food production and use. Can Earth provide enough food for an additional one to two billion people given that hundreds of millions of people still go hungry each year? Approximately one third of all food is either lost or wasted, according to the United Nations Food and Agriculture Organization, which translates to some 5% of global greenhouse gas emissions. This mountain of rotting food is set to keep growing. Absent significant policies or behavioral change, food waste is predicted to double by 2050.[19]

When it comes to solutions for the food turnaround, reducing loss and waste is arguably the lowest of low-hanging fruit (pardon the pun). And it is clear where the problem lies. In wealthy countries, fussy consumers buy more than they need and discard anything with the slightest imperfection. Retailers both train consumers to over-consume and then jump to the tune of the consumers' unsustainable and often arbitrary demands. Regulations and education can help reduce waste.

Growing more food than is consumed means that we are using land that could instead sustain greater biodiversity. Because we are, in most cases, saturating that land with chemicals along the way, we are also depleting soil and polluting land and waterways, all for naught. So reducing true waste is critical. But the unused food that remains should be directed away from landfills and used to address hunger, most importantly. It can also be used to create soil-building compost, supplement animal feed, or create energy through biogas operations. While not the highest priority for food waste, biogas can be a significant source of energy. Anaerobic digesters, which use bacteria to break down organic matter in oxygen-free reactors, can power 800 to 1,400 homes a year by turning 100 tons of food waste per day into biogas.[20]

In low-income countries, food waste is often unintentional and caused by poor storage conditions or transportation difficulties. This can be improved by better infrastructure to store, process, transport, and distribute food. New food enterprises can work with excesses in crops that ripen all at once—by turning fresh mangoes into dried mango chips, for instance. And recapturing nutrients by techniques such as biogas and composting at scale can reduce runoff and get nutrients recycled to the soil. But food wasted in one corner of the world can't really be diverted to feed the hungry in another.

Giant Leap calls for reducing food waste by 30% by 2050. Slow the pace down to just a 10% reduction by 2100, and you get the Too Little Too Late scenario.

Barriers

There are many barriers to building a resilient farming system. Simple inertia is one. Farmers are, rightly, skeptical of change. Change managed badly can be disastrous for their income. But change is essential. Take the delicious almond, for example. The vast majority of the world's supply is grown in sunny California, bringing farmers $11 billion in revenue. But almonds are thirsty, and California is running dry; it is enduring the worst megadrought in at least 1,200 years. It is likely to get worse not better in the coming decades. Precision irrigation will help, but ultimately farmers need to adapt their crops to a more suitable climate. This can be a slow process as dumping generations of wisdom is a painful business.

A second barrier is consumer behavior, which may be the biggest barrier of all. As income grows, diets change. The aspirational diet is the Western diet. Consumer demand can be managed through education and awareness campaigns; after all, most people want to eat a healthy diet. Governments can also influence consumer behavior through price (taxing sugar has reduced intake of fizzy drinks), nudging choices, and other regulations. But in democracies at least, governments are reluctant to tell citizens what they should or should not eat. Mandating a sustainable healthy diet seems an unlikely proposition.

Cost is a third barrier. Transitioning from conventional to regenerative or sustainable operations can be expensive. Farmers need help financing the transformation. A starting point is to incentivize loans for resilient farming at preferential rates and remove nonsensical subsidies that drive environmental degradation and pollution. In some parts of the world, smallholders face additional economic challenges brought on by agricultural monopolies that control seed stock or otherwise impoverish local farmers and destabilize farming communities.

As in all the other turnarounds, progress depends on fundamental changes in our political system that drives our economic operating system. The food turnaround requires introducing financial models that drive investments into the food sector and rewards those stewarding agricultural land, enhancing ecosystems, and producing safe and healthy food. This in turn will shift employment from conventional to green activities. With regenerative agriculture, for example, capturing carbon in soil instead of emitting it creates new business opportunities. Farmers given financial support to monetize carbon sequestration and the other ecosystem services will deliver value for

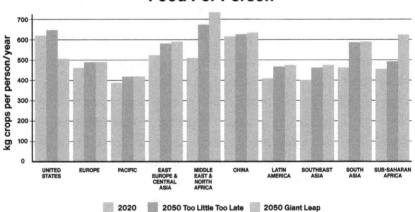

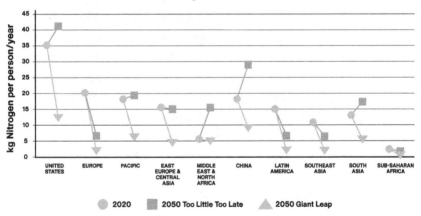

Figure 6.3. Regional crop production (*top chart*) and food footprint per person in 2020 and 2050 in Too Little Too Late and 2050 in Giant Leap (*bottom chart*). In this context, we define the food footprint (*bottom chart*) as the kilograms of nitrogen fertilizer per person per year on the vertical axis.

societies. At present, governments around the world subsidize industrial agriculture to the tune of a million dollars a minute. The money is there. It just needs to be redirected.

We all have to eat. But we should pay more for food. So-called cheap food comes at the cost of impoverished communities, chronic disease, and collapsing ecosystems. The issues around justice and inequality must be addressed. Government support should ensure affordable access to healthy diets. New legislation can ensure that even low-income families can afford good food. It can also force industrialized food producers to internalize more of the costs they have pushed off onto society—such as the costs of dealing with their pollution, their waste, and the health issues that their marketing exacerbates. That, in turn, will incentivize better corporate behavior.

A final barrier to overcome is the tangled web of regulations that incentivize vast monocultures, deforestation, and waste. To speed the transformation, governments need to revolutionize farming subsidies and tax incentives to promote sustainable, regenerative farming techniques that build on locally suitable seeds and varieties and encourage locally grown, low-carbon healthy food. Governments must also act to remove market barriers for innovative food technologies such as precision fermentation and cellular agriculture to allow new animal proteins to reach markets quickly and safely. While they are doing it, governments need to protect workers in the food industry and agriculture during the turnaround. A bare minimum is to regulate food companies to implement workers' rights across supply chains.[21] Governments must also loosen the control that agricultural monopolies have over the world's food supply, especially where it impedes farmers' rights to grow and sell food. The real challenge will be to gather democratic support for such action.

Conclusions

The UN's Food and Agricultural Organization says "Business as Usual is no longer an option." The food system is on a catastrophic pathway. Left unchecked, the Western diet will take over the world. At some point this century, we risk crossing a tipping point where over half the population on Earth is overweight or obese, while famine plagues other regions. The beneficiaries of this are the transnational corporations fattening us up. But corporations can benefit from providing healthy diets that do not cost the Earth.

The only way to lock in a turnaround of the food system is through active governments willing to build an economic system that places a value on sustainable and regenerative agricultural practices. With that will come greater national food security. The turnaround is a win-win for people and planet. But at a minimum, it will require eliminating perverse subsidies, reallocating these funds to regenerative agriculture, and tightening regulation to ban the most damaging products. It will also require making difficult decisions under great uncertainty.

For these reasons—the uncertainty, the risks to global food security, the need for active governments, and the desire for greater cooperation—we propose that governments consider creating a Food System Stability Board charged with helping to ensure food system resilience as the climate crisis deepens, pandemics become more frequent, and conflicts grow. The board could ensure short-term solutions during crises while steering long-term food system transformation. The G20 group of countries created the Financial Stability Board in the aftermath of the global financial crisis. There are indications that this has had some success in reducing systemic risk; the financial system was in a better place to deal with the shock to follow—the COVID-19 pandemic. A Food System Stability Board, overseen by the G20 or another international forum, would build on the notion of collective stability for the food system, developing more sustainable policies and regulations relating to trade, carbon storage, healthy diets, and price shocks.

The food system is beginning to transform. The signs are everywhere. Nearly one third of farms worldwide have crossed a redesign threshold and are using at least some farming techniques that fulfill our principles, including, for example, integrated pest management, conservation agriculture, integrated crop and biodiversity systems, agroforestry, irrigation management, and small or patch systems.

These turnarounds are happening on an estimated tenth of all agricultural land worldwide.[22] Perhaps this is an early sign of a tipping point to a more circular regenerative farming system. But this of course does not help dent food waste or incentivize healthy diets, and first of all, we need to ensure scale in order to drastically reduce hunger in many parts of the world—including the growing levels in high-income countries.

In summary, our challenge is to turn the global food system around so that it can securely provide nourishing and delicious foods for around nine billion people within planetary boundaries. This is genuinely possible.[23] But it means not taking any additional land or seascapes and safeguarding remaining wildlife. It also means reducing freshwater use and cutting the overuse of nitrogen and phosphorus fertilizers in rich countries, while shifting toward net positive carbon dioxide emissions without further increase in other greenhouse gases.[24] Ultimately, this means treating—and compensating—our farmers as biosphere stewards.

7

The Energy Turnaround

"Electrifying Everything"

People often express shocked disbelief that societies are failing to remove fossil fuels from the global economy at the speed and scale required. It is worth remembering that what is being demanded is a complete restructuring of the foundation of all industrial economies. Fossil fuels were at the heart of the Industrial Revolution and remain the cornerstone for economic growth away from poverty. Calls for action are entirely correct, but transformation was always going to be difficult. On top of that, the fossil fuel industry's unique position in society also means that it has become the most powerful and influential industry on Earth.

So, the fifth and final turnaround is a complete restructuring of a foundation of our economies: energy. The Paris Agreement's goal to stay well below 2°C requires approximately halving greenhouse gas emissions (at the global level) every decade from 2020, to reach close to zero in the 2050s, a trajectory called the Carbon Law[1]—not because it is mandatory (the Paris Agreement remains voluntary) but because it is required.

Within the current economic paradigm, the most important step is to increase efficiency. Like food, much of our energy is simply wasted. We throw it away. But analysis shows that global energy demand in 2050 could be up to 40% lower than today if all energy efficiencies are implemented. This could be achieved while giving all societies ample access to energy.[2]

Pushing into a new economic paradigm, a good rule of thumb is to electrify everything while simultaneously rapidly scaling renewable

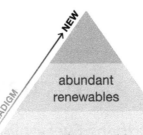

Figure 7.1. The energy turnaround starts with systemic efficiency along the entire existing energy systems. At the same time, heat, industrial processes, and transport transition to renewable electricity and energy carriers derived from it, such as green hydrogen. Large investments in abundant renewables with storage keep making power cheaper due to zero marginal costs, i.e., "free sun."

energy and energy storage to provide energy abundance. Anything that once needed fossil fuels to be burned needs to go. Gone are belching power stations, noisy polluting internal combustion engines, and inefficient boilers and heaters. In their place come roofs and fields of solar power and twirling, clean wind turbines. Electric vehicles and mass transport systems. And energy-storage solutions that range from batteries to pumped hydro, which uses renewable energy to pump water into reservoirs to ensure ample backup energy.

An important part of the transformation will be a shift to more conscious production, and to consuming less. We not only need electric vehicles, we need smaller vehicles, and fewer vehicles on the roads. Along this journey, the fossil fuel industry will put up a fight. This is why transformation will not happen without active states creating the right economic conditions for an energy upgrade. Immediate steps are to remove fossil fuel subsidies, remove market barriers for renewable energy, and make it easy for families, communities, and companies to share and trade clean energy. We also need an overall shift to circular manufacturing practices throughout our economy not only to recycle materials but also to reduce the amount of materials used in products across the board.

The good news is that the world is already on the cusp of the most profound and rapid transformation of the global energy system in history. Clean power technologies are growing exponentially everywhere. Incredibly, in 2021, wind and solar accounted for 10% of all electricity production in the world; it was just 5% in 2016. Doubling at this pace means wind and solar could account for half of all electricity supply in the early 2030s. The key issues are whether the turnaround will be rapid *enough* and whether it will be *fair*.

Challenges

The first major challenge for the energy turnaround is fairness. Carbon dioxide emissions from fossil fuels still hover around thirty-five billion tons per year. But we only get the full story when we break this number down.

Rich countries make up just a small fraction of the global population yet have been responsible for about 85% of excess global carbon dioxide emissions.[3] While some have argued that the world did not know the risks, since the start of the Industrial Revolution, over half of all carbon dioxide emissions from fossil fuels and cement have been emitted since 1990.[4] That is well after the first alarm bells were sounded by scientists in the late 1950s and early 1960s. Concern was so high by 1988 that the first Intergovernmental Panel on Climate Change was established, and in 1994, the United Nations Framework Convention on Climate Change entered into force.

The "net-zero commitments" for 2050 that are currently being made by rich countries do not account for this truly vast disparity in historical emissions, nor do they account for the fact that rich economies have essentially exported their emissions to countries like China and Vietnam that now produce the majority of their consumer goods. Not only is this unfair and unjust, it also means that the carbon emissions of a small group of rich countries will continue to grow overseas even as they scale down their emissions toward zero over the next thirty years.

There will be no energy turnaround toward a healthy planet—"just" or otherwise—if the legitimate concerns of low-income countries are not taken into account.[5] This means changing the flow of investment. But the global financial system is already rigged in favor of high-income countries, wealthy elites, and fossil fuel companies. As articulated in the poverty turnaround chapter, the global financial system needs transformation to support the energy turnaround in low-income and middle-income countries. This means de-risking investments in low-income and emerging markets that are on the receiving end of unfairly low credit ratings that generate sky-high interest rates on borrowing.

Fairness is also important in other ways. Men tend to have a bigger carbon footprint than women, for example. Ethnicity is also a factor. In the United States, Caucasian neighborhoods tend to have higher carbon footprints than African-American neighborhoods.[6] And carbon footprints tend to be tightly connected to income. The poorest people in rich countries have relatively small carbon footprints, and the billionaires in lower-income countries have very high footprints. Globally, the richest 10% have carbon footprints equivalent to the poorest 50%. The luxury carbon consumption of the ultrarich 1% accounts for 15% of global emissions. And this luxury carbon consumption is marketed as an aspirational symbol of success, vitality, and wellbeing. Without changing course, the miniscule carbon budget left for humanity's remaining time on Earth risks being exhausted out of the back end of private jets.

On top of these inequities, the fossil fuel industry has an unfair advantage. According to the International Monetary Fund, the burning of coal, oil, and gas is subsidized to the tune of $5.9 trillion annually when accounting for both direct and indirect costs, including the health costs of air pollution and the costs of climate change.[7] We need to tilt the playing field to favor cleaner alternatives. Dealing with all these challenges will require active governments willing to reshape markets (starting with removing perverse subsidies and pricing carbon fairly) and make long-term energy plans.

A final challenge is the very real risks of destabilizing societies as the energy system transforms. It is the poorest majority that are hardest hit if fossil subsidies are removed or costs of energy rise for other reasons. And predictably and understandably, these people will react against energy policies as happened to French president Emmanuel Macron. As coal industries shutter, governments will need to invest in retraining and regional redevelopment, like Spain and Germany are attempting to do. And fossil fuel companies face the real threat of "stranded assets"—pipelines, mines, and oil rigs worth trillions of dollars that will potentially become worthless if oil stays in the ground or financial capital exits the industry rapidly, with serious implications for the stability of the finance sector.

Don't Look Up

All of this perhaps helps explain why for two centuries we've looked down below ground for energy instead of looking up at the sun and the wind. This mindset needs to change. We need to bust a few myths about the upgrade to clean energy.

Myth 1: Energy transitions are slow. The transitions from biomass to coal and coal to oil have taken about sixty years. We are not starting from scratch. We are already thirty years into a renewable energy transition, and crucially, we have reached the critical inflection point in the exponential curve when the cost of renewables is comparable to the cost of fossil energy, or cheaper, in many regions. On top of this, government funding plus recent technological breakthroughs will accelerate existing trends if the right policy incentives are in place.

Myth 2: Many sectors are hard to electrify. Long-distance trucking, shipping, cement, and steel manufacturing were once considered among the sectors that were hardest to decarbonize. New solutions exist to almost entirely remove carbon from these industries while improving efficiencies.

Myth 3: It is difficult to change people's behavior. The global pandemic has shown that behavior and business models can change very rapidly and bring many benefits, for example, working from home not only reduces commuting emissions and congestion, it also helps to juggle work and family life when the right supports are in place.

Myth 4: Electric vehicles are not as good as internal combustion engines. Electric vehicles often now have higher performance in terms of speed and acceleration than fossil cars. They can be upgraded more regularly. They pollute less. They are even more reliable; an electric motor and drivetrain may have just 20 moving parts compared to up to 2,000 in an internal combustion engine, which means fewer parts to break down.

Myth 5: Intermittency means clean energy is unreliable. Many studies have shown that fluctuations in the availability of sun or wind can be offset by building more power generation capacity than needed, using energy storage systems, and creating super grids that allow energy to be transmitted over wide areas. Other factors, too, can ensure supply. And of course, nuclear energy has been proven over two generations to provide reliable electricity.

So, now some of the myths have been busted, let's explore the solutions.

Solution 1: Introduce Systemic Efficiency

No one needs steel, cement, or petrol; people need comfortable homes, offices, and other buildings, and a way to move between them. People need to do their jobs, see their friends, and access all sorts of services. In other words, what people need are the *functions* that energy and materials enable. In 2018, Arnulf Grubler and colleagues published a groundbreaking scenario of energy efficiency. They focused on end-use demand, rather than supply. What do people want to do with energy? They based their scenario on the desire for greater quality of life globally and current exponential trends in technology, for example, the move to smartphones that consolidate many services (TV, Internet, telephone, map tools) and less energy. Looking at demand for these services in both North and South, and assuming those in low-income countries want the same access to services as those in high-income countries, the team calculated that, despite a rising population and rising affluence, final energy demand in 2050 has the potential to be around 40% lower than today if this type of technology diffusion is incentivized by governments.[8] This is remarkable. The current energy narrative—that energy demand will keep rising—is not the only narrative, and it is not the most desirable outcome.

Systemic optimization toward greater efficiencies would not only save energy, it would drive down use of materials and reduce air pollution. We can find efficiencies everywhere. Most urban journeys are very short: half are under three kilometers (two miles). Large vehicles with lone drivers are not the most efficient way to move people around in congested cities. Redesigning transport systems in cities to offer cycling and walking options, efficient public transport, and shared mobility will reduce emissions and improve health without necessarily increasing commuting time. The bubble schools in chapter 2 would not be needed. Everyone, rich or poor, could breathe cleaner air in cities. In homes, improving insulation is a better solution than adding air conditioners or heaters. Refurbishing and reusing are better than demolition. Opening up buildings to daylight is smarter than running light bulbs, and so on. Whether transport, buildings, heating, or materials, the potential for radical systemic efficiency is vast across all such energy-consuming sectors.[9]

But swapping out combustion engine vehicles for electric vehicles is not the optimal solution for transport. This would still create congestion and take a large footprint to achieve. A systems-wide shift would mean that electric vehicles and the much-lauded self-driving vehicle, while they might have a place, should be *one* option among many for getting around. And don't wait for flying taxis, electric or otherwise. This does not solve congestion on the ground and creates space for a new problem down the line: more congestion in the skies. Focus on a denser infrastructure around bike lanes and public transport in walkable, livable cities.

In the exponential age we are living through, clean energy technologies will mix with other emerging technologies, and the combined effects, if directed, could drive even greater efficiencies. Again, take cars. The average European car is parked 92% of the time, often on scarce inner-city land. Car-sharing systems, using mobile phone technologies to share digital keys, have the potential to transform consumer demand from car ownership to mobility as a service. When driverless cars become mainstream, this could accelerate the shift to mobility as a service reducing the number of cars on city streets. This may be easier to imagine in California than Karachi, but the exponential technology advocates are right to remind us that the technologies that came together to give us the smartphone accelerated Internet access in low-income countries as well as the wealthy nations.

Solution 2: Electrify (Almost) Everything

The first principle of fighting the climate crisis is simple: stop lighting coal, oil, gas, and trees on fire, as soon as possible. Bill McKibben suggests we add a second rule: "definitely don't build anything *new* that connects to a flame."[10] Instead we should substitute carbon molecules with electrons wherever something needs energy.

While Solution 1 makes the case for efficiencies, our second solution is, as a rule of thumb, to electrify everything that today combusts carbon. Shift from molecules to electrons. This often makes things more efficient, as it happens. In cities, most energy demand comes from transport and buildings. The solutions are market ready: move fully from combustion engines to electric mobility and use heat pumps rather than burners for heating. The electric engine is already three to four times more efficient than fossil engines. And heat pumps are yet more efficient compared with fossil heating. The more you electrify, the less (primary) energy demand you get. This is a general rule, and there is no simple electrical fix in all sectors, for example, steel production or shipping. But green hydrogen and ammonia can replace fossil fuels in these sectors, and major steel, fertilizer, and shipping transnational corporations are already committing to these solutions. Still, substantial governmental support is urgently needed to get the ball rolling, given that the upfront cost of renewable energy still is much higher than that of conventional fuels.

We can leverage the scaling potential of solar, wind, and batteries to transition industries onto the new clean energy system. The dividend here is the opportunity to rapidly bring down operating costs in the long run—in a system where clean energy is being produced with nearly zero marginal costs for most of the year. In summary, we are facing a temporary cost hurdle, once it is passed (with the associated upfront investment costs), we will be closer to an electric wonderland with lower costs per energy unit.

But we have to acknowledge risks in this transformation. How can it occur without human exploitation in new mining operations? And how can it roll out without the destructive expansion of and pollution from mining and other extractive industries? Compensation and justice must be embedded and regulated into the shift to the new technologies.

Solution 3: Exponential Growth in New Renewables

The energy paradigm shift calls for the almost complete replacement of the fossil fuel industry with clean green energy. In many parts of the world, renewables are now the cheapest source of new power generation—if you look at the lifetime cost of energy. They have reached market maturity and will outcompete fossil fuel incumbents on price and performance, while drastically cutting pollution.[11] But only if the government provides the necessary subsidies, so the upfront investment cost is brought down to that of the fossil alternative. Luckily, this is happening. The share of wind and solar in the energy mix has been doubling globally every five years. An ever-longer line of energy experts are pointing out that exponential technological change is finally here.[12]

Renewable energy technologies keep getting cheaper because they follow learning curves. For each doubling of the total installed capacity, their cost declines by around 20% to 25%. But power from fossil fuel technologies do *not* have such learning curves. The technology behind coal-fired power stations has barely changed in decades, and this will not change because these are mature technologies. Another reason for the difference in innovation speed is that fossil fuel infrastructure and plants are often big and bulky rather than small and granular. Small granular technologies, think smartphones or electric vehicles, have rapid innovation and marketing cycles. Wind and solar do, too. Therefore, renewables will plausibly keep getting cheaper than fossils in the coming decades, outcompeting them in ever more applications and areas.

Exponential change is not just about adding ever more solar panels or wind turbines. It's about how a number of digital and granular technologies interact in systemic, self-reinforcing ways and what comes from this. It is feasible to meet 100% of electricity demand in most regions of the world using solar, wind, and battery (SWB) solutions, given parallel investment in smart grids and super grids (wide-area transmission networks). That means solar panels everywhere on existing infrastructure, integrated with transmission and energy storage. That storage would come from a variety of methods: chemical and gravity-based batteries, pumped hydro, thermal, compressed air, or combinations of all.

Significant challenges remain, such as sourcing the additional metals and materials and financing a build-out. While no realistic industrial changes predicated on maintaining human civilization can sidestep this hurdle, this can certainly be achieved without injustice and exploitation.[13] But not without government support to accelerate the process, so things happen before the world gets too warm.

Before we leave our discussion on the solutions and move to the barriers, we'd also like to explore the potential of energy super-abundance. The exciting spin in the tail from the acceleration in new renewables is what results from *overbuilding* the new energy supply and network to go beyond current demands. As the costs of solar, wind, and batteries fall, we reach a point of clean energy super-abundance at near-zero marginal costs. Rather than concern about intermittent supply, the implication is that the clean energy disruption based on solar, wind, and batteries heralds the potential to break through to a new energy system the likes of which we have never seen before. It will enable humanity not only to meet our current energy needs sustainably but to electrify for a vast array of other things that are economically impossible within the current system. We can power carbon and capture and storage systems like direct air capture to assist with the drawdown of excess atmospheric carbon dioxide. This is needed to go beyond net zero and create a climate-positive energy system that helps reduce carbon dioxide concentrations toward pre-industrial levels.

As Earth for All Transformational Economics Commissioner Nafeez Ahmed points out, the electrification of a vast array of industries and sectors, from wastewater treatment and desalination to recycling, from mining to manufacturing, will reduce their primary energy consumption and shift to clean cheap power. This can also bring about a nearly fully circular economy as there is sufficient power to purify and upcycle the waste fractions. For the first time, the vast amount of additional power generated by the new system, combined with efficiencies, will allow us to cleanly sustain the extensive new industrial processes required for the circular economy in a way that was previously inconceivable.[14]

The Energy Turnaround in the Earth4All Analysis

How far and quickly can the three solutions push the energy turnaround globally? It is clear that the speed of renewable deployment will determine how fast fossil fuels can be phased out. Figures 7.2 through 7.5 show the effect of going from no extra action to a strong turnaround, by ramping up investments in systemic efficiency, electrifying everything, and making renewable energy abundant. Global energy costs (total annual costs in both investments and operations) are higher in Giant Leap than in the Too Little Too Late scenario for the 2025–2050 period. But global CO_2 emissions from fossil power production then decline to zero around 2050, while annual total energy costs become much lower from around 2050 on, as the energy system by then has a huge renewable capacity driven by free sun and wind.

Barriers

It is clear the energy turnaround is already underway. The biggest obstacles are no longer technologies or inadequacies of solar, wind, and other clean energies. Most major economies have now committed to net-zero emissions by 2050, or in the case of China and India, 2060 and 2070 respectively. As the upgrade, and in some cases complete transformation, gathers pace, it is possible to imagine

Renewable Energy

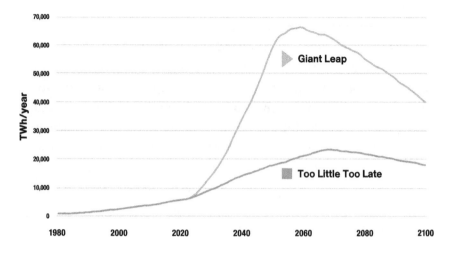

Figure 7.2. Renewable energy production with storage soars globally in the Giant Leap scenario, helping to electrify "everything" and making clean power available for all humans. Source: E4A-global-220501.

CO$_2$ from Energy & Industry Production

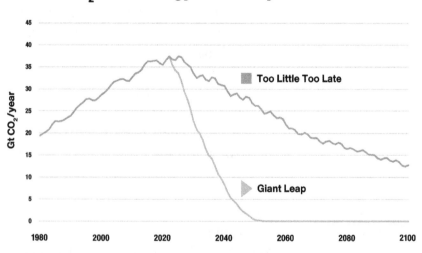

Figure 7.3. Global CO₂ emissions from energy and industrial processes decline rapidly in the Giant Leap scenario, following the Carbon Law and making it possible to keep temperatures below 2°C by 2100. Source: E4A -global-220501.

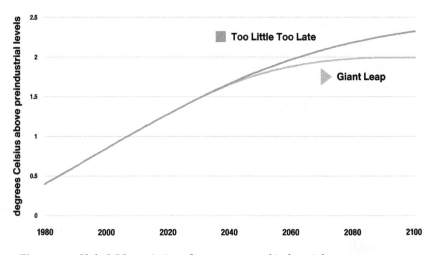

Global Warming

Figure 7.4. Global CO_2 emissions from energy and industrial processes decline rapidly in the Giant Leap scenario, following the Carbon Law and making it possible to keep temperatures well below 2°C by 2100. Source: E4A -global-220501.

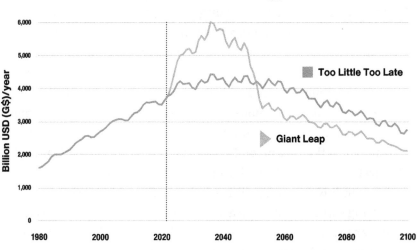

Total Cost of Energy

Figure 7.5. Turning energy investments around from conventional fossils to the solutions in the Giant Leap scenario leads to higher energy cost in the first decades compared to the Too Little Too Late scenario, but dramatically lower cost in the long run. Units: billion USD, at PPP-2017 constant prices. Source: E4A-global-220501.

even these countries achieve their targets earlier than expected. But despite rhetoric, most governments have failed to commit to halving emissions by around 2030, preferring to set more distant goals far over their political horizons, keeping us on the Too Little Too Late scenario. This is fully understandable given the huge subsidies needed to bring down upfront investment costs that need higher taxes or government debt to finance. But these subsidies are paltry in a long-term global perspective to deliver energy security and planetary stability.

A second barrier are the enormous subsidies received by the fossil fuel industry, and their lack of accountability for the damage caused. There is little logic to keep most subsidies (some though are designed to help people on low incomes access energy, and these must be redesigned).

A third barrier has been to implement a fair price on carbon emissions. For decades, climate and energy politics have been dominated by the idea of pricing carbon emissions. Carbon pricing provides an elegant response to a complex problem: increase the cost of releasing carbon into the atmosphere and let energy markets take care of the rest. While the policy may look good on paper from conventional economic thinking, in practice it has proven weak in the real world. Since the beginning of most carbon trading schemes, the annual supply of pollution permits has most of the time been consistently higher than overall pollution. And after thirty years of policy on the issue, only 20% of global greenhouse gas emissions were covered by some carbon pricing initiatives in 2020. But of these 20%, less than 5% (meaning less than 1% of total global emissions) are currently priced at a level consistent with achieving the temperature goals of the Paris Agreement.[15] New thinking is needed to price carbon fairly.

Pricing carbon is linked to a fourth barrier. In many democracies, politicians with ambitious energy agendas have so far struggled to reach power at the national level (though it is often a different story at the state and city levels). This is despite a plethora of surveys indicating public support for government action to deal with the crisis. The political problems are, of course, that citizens in general oppose more expensive gas or electricity bills, and that energy poverty is a real concern in most countries. Thus, there have been severe political constraints on climate action. Discontent is often rooted not in doubts about the need for climate action but in a distaste for solutions that put too much of the burden on the lower-income groups. Citizens distrust politicians, bureaucrats, and elites and feel that the ruling class treats them with contempt.[16] From the French Yellow Vests to demonstrations in Iran, Turkey, Nigeria, Mexico, Jordan, or Kazakhstan, the opposition has been predictable. In a nutshell, "Axe the tax!" slogans have worked wonders for politicians.

This is why a fee and dividend approach, mentioned in previous chapters, merits further discussion. In the next chapter, we will discuss the full concept of a fee and dividend and how it applies to all global commons—that is, all the natural resources that are at risk of destabilizing as a result of human pressures. But for now, let's stick with the atmosphere and carbon. We know that the rich consume most of everything. Charging a fee for carbon emissions and then distributing all fees back to every citizen equally is fair and reflects the principle that we are all stewards of the atmospheric global commons. In addition, it has the effect of reducing inequality because heavy users of carbon must pay more. And those that use little receive compensation. A surprisingly large number of economists agree it is a good idea. According to the *Wall Street Journal*, in 2019, 3,500 economists endorsed a carbon fee and dividend approach to pricing carbon as "the most cost-effective lever to reduce carbon emissions at the scale and speed that is necessary."[17] The fee and dividend approach also keeps the revenue raised out of the general tax fund. The fees are only for redistribution, not for diverting to a thousand other policy objectives because the heart of the matter is trust. Clarity is critical. It is important that people see where the money is going and that people benefit personally and visibly from changing their behavior.

Scaling all solutions until we reach net zero by 2050 is not only extremely ambitious but also impossible to accomplish *unless* we can overcome the barriers mentioned above: the dramatic inequalities in footprints and energy access between high- and low-income countries on the one hand and within-country political constraints on the other.

This is why the practical and technical solutions above need coordination by trusted and active governments to succeed with reducing both emissions and inequalities at the same time. This means recognizing and redressing existing global inequalities. The world's largest economies—the US, EU, and China—must ramp up (at least triple) annual domestic investments in renewable capacity at high speed. Together these three emit roughly half of all greenhouse gas emissions from their own territories. In addition, to speed the turnaround in low-income countries where most of the world lives, there is the need to ramp up action in three key international areas:

- Redress carbon footprint inequalities by a massive increase in climate finance.
- New financial architecture for debt resolution and to incentivize green investments.
- Reforms of Special Drawing Rights (SDRs) and trade rules to enable green economic trajectories.

Expanding on the first point, a massive injection of climate finance is urgently required. Obviously, there is no excuse for the rich countries to renege on their promise from 2012 of $100 billion per year in climate finance. They have consistently underdelivered, and the funds should be paid in full covering all the shortfall. One immediate and costless means of providing this is to expand SDRs, the international liquidity created by the International Monetary Fund. A much larger issuance of at least $2 to $3 trillion per year is needed, and is easily achievable, earmarked for investments in clean energy systems. In addition, there is a strong case for high-income nations to recycle their own allocations of SDRs (which they are unlikely to use anyway) to regional multilateral development banks for such investment.[18]

Second, create an international framework for sovereign debt resolution that dramatically reduces the unpayable debt burdens in low-income (<$10,000 per person per year) countries. This restructuring mechanism should necessarily include not just bilateral and multilateral lenders but also private lenders by mandatory regulatory and legal changes. In addition, there must be stricter regulation on private financial markets to prevent any more "brown" and carbon-intensive investments by private lenders and bond holders, and to incentivize green investment.

Third, the concentration of the necessary knowledge and technology in the global North companies is deeply dangerous for all. It prevents the dissemination of critical clean technologies. The global

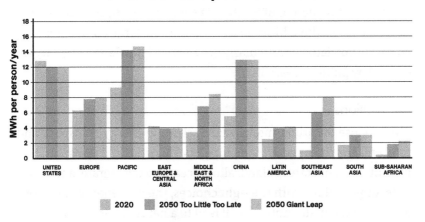

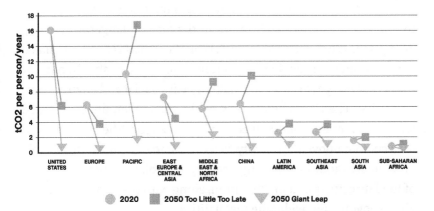

Figure 7.6. Large differences in regional energy footprints: showing the power consumption and CO_2 emissions per person for 10 regions in 2020, 2050 in Too Little Too Late and 2050 in Giant Leap. Source: E4A-regional -220427.

system of intellectual property rights put in place by the World Trade Organization must end for critical technologies that are essential for leapfrogging toward a green transition in low-income countries. Far from encouraging more invention and innovation, it has led to monopolies of knowledge and "rent seeking" at the cost of the public good. It is preventing the possibility of ensuring that all countries can access the crucial technologies, from vaccines to solar. When low-income countries seek to encourage renewables through subsidies to their own producers, they quickly face cases in the WTO. Avoiding the systematic destruction of Earth's life support system should perhaps not depend on the whims and profit of a few large companies that control knowledge, particularly when that knowledge was largely created by public research.

Conclusions

The collapse of carbon-intensive industries in energy, transport, and food will end the huge demand for global logistics and transport, free up billions of hectares of land, allow oceans to regenerate, and eliminate air pollution. With the right choices, the new energy, transport, and food system disruptions will lead to a net reduction in the materials intensity of human civilization. And provide sufficient energy to the poor.

While the ultimate demise of the age of fossil fuels is unstoppable, the survival of human civilization in the face of dangerous climate disruptions is not. It depends entirely on the societal choices we make today. Whatever specific tools nations choose, they should aim for those that merge systemic productivity with social and environmental justice. The right choices now could open up an unprecedented era of clean energy abundance by the late 2030s. The new possibilities could empower humanity to solve some of its most intractable problems: in addition to energy scarcity and volatility, the persistence of food insecurity and malnutrition, along with the entrenchment of global poverty and widening inequalities.

8

From "Winner Take All" Capitalism to Earth4All Economies

A New Economic Operating System

So there we have it: five extraordinary turnarounds that can transport us into the next decades of the twenty-first century far more safely and comfortably than our current course. If you feel the ambitions are daunting, you are right. If you suspect we could never accomplish them, think again. Getting humanity back within a safe operating space in this century may be complex and monumental, but like many other complex and monumental undertakings, it can be set in motion by a handful of well-chosen levers, by groups of committed people.

Those levers are in plain sight and waiting to be pulled. And they all reside in one sector: the economy. You may have noticed them embedded in all five turnarounds discussed in the preceding chapters. Key among them:

- Creation of Citizens Funds to distribute the wealth of the global commons fairly to all citizens.
- Government intervention (subsidies, incentives, and regulations) to accelerate the turnarounds.
- Transformation of the international financial system to facilitate rapid poverty alleviation in Most of the World.
- De-risking investments in low-income countries and cancel debt.
- Investment in efficient, regenerative food and renewable energy systems.

Traditional economists will take pause here, rightly recognizing these shifts as catalysts for massive economic transformation. Some will also, undoubtedly, fear these shifts will lead to an abrupt end of economic growth and, then, to economic collapse. On these counts, though, they would be wrong. To understand why, it will help to understand a little bit about leverage points and why they often surprise us.

Donella Meadows, the lead author of *The Limits of Growth*, famously described leverage points as "places within a complex system (a corporation, an economy, a living body, a city, an ecosystem) where a small shift in one thing can produce big changes in everything." Seems simple enough, but she also noted another reality: While people often intuitively know where leverage points lie, they tend to push them in the wrong direction, creating a web of unintended consequences. That is, said Meadows, exactly what has happened with growth: "The world's leaders are correctly fixated on economic growth as the answer to virtually all problems, but they're pushing with all their might in the wrong direction."[1]

Hence, we end up with global economic policy that was crafted to alleviate poverty but many decades later has morphed into a poverty trap, economically enslaving whole nations, destabilizing democracies, and "crowdfunding" environmental catastrophe. We have watched the purpose of our economy morph from valuing our future to discounting it entirely.

No surprise, then, that many increasingly angry citizens have intuited that conventional economic thinking is no longer delivering economic security, and agency, for them or their families. And no surprise that we need to upend that conventional economic thinking. But does this necessarily mean an abrupt end to growth or the risk of economic collapse? No and no.

The energy transformation alone will drive economic growth. How can it not? It is nothing less than a complete reorganization of the foundation of the economic system. It will create economic optimism, investment opportunities, and jobs in all sectors. And if the transformation is managed fairly, where everyone is granted a stake in the future, this will help ensure the necessary political stability to avert the risk of economic collapse. That said, we should be largely agnostic about growth—it depends on what is growing. For sure, economies need to shrink their material footprints by shifting to circular models. And ultimately, the economic focus has to shift to growth in wellbeing. This is beginning to happen. Some local and national governments are experimenting with new economic models. We've already mentioned the Wellbeing Alliance of New Zealand, Finland, Iceland, Scotland, and Wales, but also cities like Amsterdam, Brussels and Copenhagen are actively challenging the old values of their economies and finding ways to turn them around.

But is the cost of avoiding catastrophes prohibitively expensive? Perhaps this is the reason for the hesitancy? Well, first of all, it is not a cost. It is an investment in the future. We estimate the investment needed is roughly 2% to 4% of global income per year. The largest investments are needed in sustainable energy and food security. This estimate aligns well with other studies.[2] Indeed, the writer and academic Yuval Noah Harari and his team have pored over various economic and climate reports and found that estimates for the energy turnaround converged around 2% to 3% of annual global GDP. As a comparison, governments directed the equivalent of over 10% of global output to counter the shock of the pandemic.

If the benefits are so great and the investments so small, relatively, what is holding us back? Ultimately, our mindset; the all-pervasive winner-takes-all worldview.

The Rise of Rentier Capitalism

Our economy has undergone massive transformations before, particularly since World War II. What came with that transformation was a gradual shift in mindset—one that led us away from economies organized, however imperfectly, to serve the public good and, since 1980, toward a dominant economy that served a small set of global elites. The wealth of this "rentier" class grows from the ownership of financial assets. In systems terms, a vicious cycle kicked in: As assets begat more assets, they concentrated in fewer and fewer hands. We can see how this shift unfolded through three economic narratives—specifically the narratives of wealthy nations, because it is their metamorphosis that dramatically changed the economic landscape worldwide.

Narrative one was dominant in the West during the postwar era (1945–1975). Economies were more national than global. Decision-making was a three-way split between business, organized labor, and government. The banking and finance sector served a secondary role; it was a support for the overall economy, not its driver. Key aims included full employment that would then support social safety nets. Infrastructure was supplied by the government, and taxation was on profits, income, and consumption. This economic system drove stability, prosperity, and greater equality in some parts of the world. Some challenges emerged over time: inflation, competition from new industrial nations, labor unrest.

What followed? Narrative two, the market liberalization era (circa 1980–2008). Dominant Western nations embraced globalization in return for efficiency. Government functions were increasingly privatized. The power of government and organized labor weakened while the power of business expanded. The finance sector rose to dominate the economy, becoming highly deregulated, expanding globally. Government priorities shifted at home to helping the market work well, subduing inflation, and limiting their own direct economic activity. Taxes on profits and capital were lowered. New problems emerged: expanded private debt, weakened infrastructure, short-term financial decision-making, and rising inequality.

By the time of the 2008 financial crash, it was clear that the second narrative had been tested as a social contract between citizens and governments and failed. What was revealed in that moment of crisis was the identity of most governments' first priority: to protect asset prices and the financial system. Emphasis turned to the supply of new liquidity, the forcing down of interest rates, and the purchase of shaky assets. Even worse, the costs of the bailout were transferred to the public purse.

After 2008, attempts were made to reboot the second narrative using debt while imposing austerity on the public sector. Ongoing structural economic forces accelerated inequality, expanded the numbers of the economically insecure, shrank the middle classes, and undermined growth. Perhaps it is no wonder that by 2015/16 different shades of populism were rapidly gaining ground, particularly in the English-speaking world. Yet come the COVID-19 pandemic, trillions of new dollars flowed into the financial system as a priority. Rinse and repeat.

Narrative three, then, has been the steady rise of a parasitic rentier economy in the name of free markets. Gone is the economy most people think we have—one organized around production, consumption, and exchange. Money is made on money and the shifting value of various assets from stocks and bonds to real estate to intellectual property and crypto. The manipulation of these financial assets now dominates economic decision-making across the globe.

This "unsustainable monopoly game," says Dr. Mamphela Ramphele, "needs to be unmasked for what it is—a self-serving platform for those who have gamed the system as both players and referees."[3]

Indeed, the shift to rentier capitalism has cost billions of people opportunity, security, and wellbeing. The toll it takes on social and environmental justice is becoming ever clearer. As it does, calls for a new economic logic have grown louder.

Rethinking the Commons in the Anthropocene

Trace economies back further, or look more deeply in the modern world for alternative models, and you will find an economic organizing tool that is the polar opposite from today's rentier capitalism. It rests on the original narrative—one that focuses on securing a people's wellbeing by securing their shared commons.

A simple example. Villagers far up in Nepal's valleys used to help maintain canals downstream while villagers downstream helped maintain dams upstream. In this way they all got access to a common resource—water. They got a dividend from the systemic benefits, and they pooled efforts into maintaining these cultural and natural assets.

The world was once replete with sophisticated examples of commons management. These systems ensure that land remains a public good to be held in stewardship by current generations for the benefit of future generations.[4] In their truest form, they are still found in Indigenous and traditional cultures. The notion of the "commons" runs like a thread through human history. That thread mostly records displacement and dispossession, but it never disappears. In a twenty-first-century revival, it is showing up almost everywhere, including in a suburb of London under a tree.

Ankerwycke's Yew tree, in Surrey, England, is about 2,500 years old. It was already old when England's Charter of the Forests was sealed in 1217, recognizing the inalienable right of local people to meet their needs for subsistence from the various fruits of the forest, from the collection of firewood or peat, fallen wood, the seasonal grazing of animals, and more. In short, to access their common inheritance on land then owned by others in a feudal system. It was actually a smart move. The king did not want to risk food riots, banditry, and a population unable to pay taxes. Nor did he want to give the major landowners leverage in a power struggle.

The charter was read out four times a year in every church in the land by locals courageously reminding the powers that be that the "commons exist for a way of living, not profits." History tells a different story. Nearly 400 Acts of Parliament between 1760 and 1870 confiscated nearly 2.8 million hectares (corresponding to one fifth of England). Strangely, the charter was only repealed 754 years later, in 1971, as neoliberal economics began to crystallize. The Ankerwycke Yew still stands, though, and in 2017, on the charter's 800th anniversary, people gathered there to issue a different reminder—that reviving the commons is about valuing our future.

Land was only the first target of an assault on common inheritance. Eventually the obliteration of user-oriented customs and practices turned global. Confiscation, colonialism, and slavery replaced a degree of self-reliance, and the misery and uncertainty of exchanging labor for wages displaced independent livelihoods. Property rights became enshrined ahead of human rights, a tension that is pervasive across all times, including our own.

The excesses of early industrial capitalism did not go unchallenged. Social evolution—and indeed revolutions—from the end of the eighteenth century led to pressure first to abolish slavery. The emerging labor movement after the political convulsions of the mid-nineteenth century sought to exert a counterforce. By the turn of the twentieth century in disparate parts of the world, a socialist counterpoint—state control of key resources in the name of the people—acted as a political challenge, another polestar. In the West, especially after World War I, the piecemeal establishment of a public sector replaced the notion of the commons with the idea of "welfare": citizens would be working full-time but paying taxes toward pensions and unemployment benefits.

Public investment would then, in this social contract, provide essential social goods and services at low cost: education, transport infrastructure, health, open spaces, and housing. Businesses would contribute tax income to this public sector—in compensation for exclusive enclosed access to the Earth's endowment. The idea of the commons did not die, but it was submerged. The tendency to encroach on this planetary endowment and that bequeathed by previous societies continued unabated.

Even knowledge was enclosed, through the expansion of copyright and patents—in other words, intellectual property rights. Increasingly, by the late twentieth century, this appropriation of the commons was extended into the digital arena—capturing the value from personal data, for example—as well as into the biological world. Private claims were made on seeds and organisms all over the world, abusing the genetic commons that are necessary for long-term environmental integrity.

Eventually money itself, that practical public utility, shifted. Credit was now mainly created by a loosely regulated, private banking system, not by governments. This credit was increasingly directed to what banking perceived as a priority, and overwhelmingly banks chose to lend money for the purchase of existing assets. Even cash, that minimal guarantee of independent economic action and exchange, is now in steep decline too. In sum, large parts of what used to be commons, available to all, have been enclosed by corporations.

Along the way, citizens of nations that were resource-rich but cash-poor watched their commons get absorbed and exploited as mining, rivers, timber, and other extraction rights were sold away, or possessed in loan repayment schemes. Vast swathes of agricultural lands were swept up into the hands of those who lived thousands of miles away, and even human capital was appropriated as regional economic systems were swept away by globalism.

The challenge now is to rebuild an economic operating system that values the commons and operates in a twenty-first-century context. Dr. Ramphele puts it this way: "There is a need to become Indigenous again with community-based economic systems based on local production and consumption needs. Reciprocal exchanges of surpluses with neighboring communities to ensure wellbeing for all and protection of ecosystems need to be revived. The model of inter-linked village communities that restores a sense of place and belonging to all people."

Doing that requires a major change in the economic gameboard.

The Conventional Economic Gameboard

Rentier capitalism is not the only economic system in play in the world today. China operates a state capitalism model in which the control of land, the money system, and significant core businesses (steel, cement, rail, utilities, and the application of credit) have been retained by the central government, often through state enterprises. China has a very different compass, evidenced by moves in 2021 to damp down on property bubbles, take down tech moguls that were too outspoken and rich, and create a less unequal society under the leader's mantra of "creating common prosperity."[5]

However, there seem to be enough similarities between the systems to describe a couple of significant stages within each of these modern economies. At first, develop markets at home, create credit, invest in infrastructure, and support the growth of an industrial economy, as opposed to an agricultural economy. As productivity and opportunities expand, urbanization concentrates the population. The scale and efficiency of production increases, and unless other markets become available (for example through colonial expansion or capture of market share elsewhere), it soon falls into overproduction and margins decline. The workforce moves from the primary sector via the industrial secondary sector, and then to the tertiary service sector.

Industry at scale is just too efficient at what it does (if utterly negligent in accounting for social and environmental impacts). Wages and incomes first rise (so poverty declines), and domestic demand rises too. But so do savings. Investment bodies begin to look elsewhere for returns, as do savers. If the productive economy has slower returns on investment, then the second stage sees shifts toward existing financial assets, particularly real estate, stocks, and bonds. At the far end comes currency and commodity speculation with complex derivatives and futures instruments.

An expanded financial sector also drives down the number of competitors in a sector. Private equity firms roll up ownership of small companies in niche sectors. This results in more monopolies and the phenomenon of tech companies "blitz-scaling" to monopolize new sectors at all costs. So the system functions in the very opposite way to effective competition in a free and fair market. The rentier economy is about charging for access to resources, limiting competition, and, in short, extracting value more than creating it. It then funnels such economic rents (unearned surplus) to the benefit of what economist Michael Hudson calls the FIRE sector: finance, insurance, and real estate. This activity is not a bug in the current system, it is a predictable structural outcome of the rentier capitalism gameboard.

Along the way, workers become expensive to hire, as they accumulate higher overheads. Access to mortgages, to credit, and the associated debt service costs are higher than they need be, and that eats up larger shares of household budgets over time. Workers are replaced with more robots and artificial intelligence. This diminishes the overall quality of life as workers' share of national income falls, as it has been since 1980. Gig work, welfare erosion, inequality, and insecurity abound. Add in the increasingly disrupted ecology—climate breakdown, pandemics, biodiversity loss—and rent-seeking brings societal collapse closer, as our Social Tension Index shows, as ever-fewer people can see how the system can be reset in their favor and in time.

But if we know something of the economic gameboard and take a look at some of the systemic tools available, it is surely possible to imagine and then actualize the five extraordinary turnarounds that are the focus of this book in a way that is about the common good.

Take a look at the economic gameboard depicted in figure 8.1, which reflects the flows in the Earth4All model. Bang in the middle are two main players that represent the economy as introductory economics textbooks and most people see it: producers (such as firms) in a market exchange with consumers (households, workers, citi-

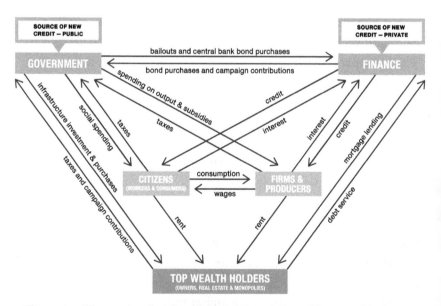

Figure 8.1. Like gravity, the current economic gameboard has a trickle-down effect—but the money flows down mainly to the wealthy. "Financial wealth accounted for $250 trillion (52%) of global wealth in 2020. But real assets effectively doubled the size of the pool. Led primarily by real estate owner-ship, these assets generated $235 trillion, or 48% of total global wealth." Anna Zakrzewski et al., *When Clients Take the Lead: Global Wealth 2021*, BCG (June 2021). See also Sean Ross, "Financial Services: Sizing the Sector in the Global Economy," Investopedia, September 30, 2021.

zens). In a rentier economy, we also have two other major players: the finance and banking sector and the owners of assets, real estate, and monopolistic arrangements. Government tries to oversee this, but has been weakened and cornered by neoliberal ideology.

The gameboard shows where money comes from and how it travels through the economy, before it trickles down—not to the many poor but to the wealthy. So, let's look first at the board's two "money-making trees." At the top right is the private moneymaking tree. The "secret" of banking is that money is mostly created out of nothing, as credit each time a new loan is made. Here moneymaking is performed by private banks, which create money as credit, with debt as its dark shadow. It feeds the finance sector.

On the left top is the public moneymaking tree. It is available to any government with a sovereign economy, but it is noticeably absent from many countries with a weak or non-sovereign currency, as explained in chapter 3, on the poverty turnaround.

The finance sector, top right, today currently dominates the game-board. The second most dominant sector is comprised of the top wealth holders—the owners of monopolies and existing assets such as real estate, intellectual property, and mineral resources. They are in an intimate alliance with the creators of credit.

In today's economy, these two dominant players always win; they have become the beasts the rest of the players must constantly feed. They are "too big to fail." Money flows to the few top wealth holders. It is worth remembering, though, that money is after all a social con-struct. Finance and banking licenses are granted and (at least nation-ally) regulated by governments. Which means the game doesn't always have to play out as it currently does.

Redrawing the Gameboard

Now let's imagine what the gameboard could look like if we pull three levers.

The first lever—a Citizens Fund to distribute universal basic divi-dends generated from fees on wealth extraction and use of shared commons—adds another money tree to the board: nature. It was there all along—the invisible source of all wealth—yet it wasn't valued, which made it easy to ignore, and to destroy. Once the benefits from nature are shared with citizens, as in the commons of old, wealth begins shifting back to workers, communities, and households.

The second lever—regulating finance to invest in strategies that address inequality, climate change, and other crises—shakes the private money tree in new ways. Regulations can channel lending away from fossil fuels and unsustainable agriculture and toward clean energy and regenerative practices. Or away from luxury apartment complexes and toward affordable, resilient community-focused buildings. To make this happen, two players wake up: The government takes a stronger role to encourage the shift, and citizens start viewing themselves as a public whose future is worth investing in.

To stimulate both these levers, governments with sovereign currency can shake their public money tree and reap rewards that translate into long-term environmental and human security. Remember, governments that fully control their own currency, and have a currency that is not backed by a commodity such as gold, don't have to spend only what they earn or borrow. They can actually spend money into existence, as long as there is unused real capacity in the economy, without causing excessive inflation.

The third lever—the cancellation of unfair debt—upends the gameboard dramatically. Here, governments step in to insist that debt held by creditors on unfair terms be forgiven. Cancelling $900 billion in international debt burdening low-income nations could free up spending that addresses post-COVID poverty and replenishes or sustains resources.[6] Such a move would affect almost a billion people.

Now, assets on our gameboard no longer concentrate only in the financial sector and with owners. They begin to filter back to the middle of the gameboard—producers, consumers, governments— and they support the formerly invisible foundations in drastic need of investment: nature and society, or put in modern economic terms, natural and social capital.

In this context, economic growth, that holy grail of the modern economy, takes on an entirely new character and purpose. We shift

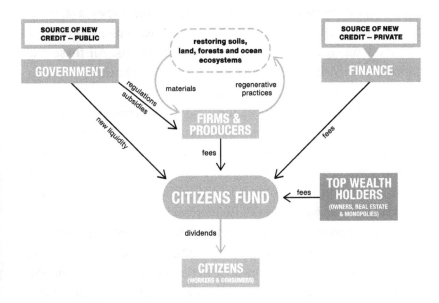

Figure 8.2. A fairer trickle down of wealth via the Citizen Fund to all citizens. Those who extract wealth from common resources (productive, natural, intellectual, social commons) are charged fees that go into a Citizens Fund. This mechanism corrects and counters the unfair neoliberal gameboard dynamics in figure 8.1 and provides essential safety nets for citizens during periods of economic transformation.

our focus away from measuring progress by the narrowly conceived annual flows of consumption and production, and toward measuring growth in a broad, shared wealth. Wealth in this sense needs explaining.

The productive commons include those human-made assets that are readily measured and accounted for in money and market prices: machines, roads, Internet, power grids, water, harbors, patents, all kinds of built infrastructure with public access, and above all, an educated, capable workforce. The natural commons consist of land, soils, a stable climate, rivers, coastal seas, deep oceans, seaweed and kelp, forests, clouds, mountain ecosystems, mineral deposits, a safe ozone layer, and the Earth's other life-supporting systems. The most important form of natural capital is the capacity of intact ecosystems to create conditions conducive to life, and to repair themselves. The social commons include arts and culture, shared knowledge, traditions, laws, databases, social media data, genes, open-source algorithms, languages, norms, and shared worldviews. All these and more influence the stock of interpersonal and institutional trust, the core of social capital.

In other words, if the stocks and flows in the economic system are managed to prioritize financial capital, you will see GDP and national wealth rise but environmental stability and social wellbeing erode. That is how the system is playing out on the conventional gameboard.

If production is then harmonized with natural stocks and the materials produced use and reuse increasingly fewer resources, you may then experience lower ecological footprints and genuine green growth. This is given that the rate of change in resource productivity is fast enough to reverse resource depletion or climate emissions. But not until you also account for human wellbeing through the fair distribution of assets and opportunities, and the maintenance of social capital, do you arrive at a truly healthy economy.[7]

In a healthy economy, what many refer to as a wellbeing economy (see box page 29, What Is Wellbeing?), future prosperity relies not as much on the annual ups or downs of economic activities, which is what GDP per year measures, but on how well those activities build and maintain the commons—in other words, all of its broad capital stocks—over time. The change in a nation's wealth can be measured as the annual change of all the capitals in a balanced way over time.

In this approach to running an economy, emphasis is placed on balanced growth in the broad wealth that serves all people, no longer in just growing incomes and the wealth of the top wealth holders. It is the kind of wealth that truly trickled down to Shu, Samiha, Ayotola, and Carla as we imagined their paths through the Giant Leap future in chapter 2. As they grew up, these girls would have received endowments from Citizens Funds for their share in the wealth of the global commons. Those endowments provided them a positive start in life, healthy diets, and better access to good health and education. Later on, the funds gave them economic security as their cities transformed. They allowed them to retrain as some industries contracted and others grew. Unlike their parents' generation, throughout their lives, these four felt governments were working in their interests much of the time.

Short-termism: The Road to a Parasitic Financial System

So why do we so tenaciously cling to our current economic system? Even if we know it leads to value extraction from natural and social capitals? And even if we know it neglects long-term value creation through reinvestment in the commons? The reason is not (or at least mostly not) evil intentions. Rather, the traditional and systemic focus has been on short-term profit flows and decision-making at the expense of long-term commons stocks and survival and planning for resilience to future shocks and stresses. Most economic managers engage in short-termism because they tend to be evaluated on their one-, three-, and five-year performance records. Further, most investment fund managers have (incorrectly) assumed that the consequences of the climate crisis were far in the future.[8]

A frequently remarked feature of the last twenty-plus years is that interest rates have been pushed down through the operation of central banks. They've been driven by the fear that a sudden uptick in interest rates would send shock waves through the precarious mountains of debt and cause a sharp increase in business failures and deep recession. Low interest rates were intended to encourage borrowing for investments in real productive capacity. More often, though, they have encouraged borrowing for yet more "paper" asset purchases and to increase the search for capital gain. The low rates also herald a rather simple assumption: Since the cost of borrowing is now so low, governments with a sovereign currency can easily borrow what they want to—and they did so in 2008 and again in 2020—and without "burdening" the future with interest on the debt.[9]

This underlying dynamic is what drives a systems failure. The overly financialized economic system becomes parasitic, extracting more out of the commons than can be regenerated and undermining the broad wealth that supports human security. When inequality gets worse, wellbeing and trust is undermined, and social tensions rise.

Putting the Systems Change into Effect

So now we understand that governments can adjust the stocks and flows on the gameboard—if they want to activate the five extraordinary turnarounds we describe in this book—by creating the vast sums needed to implement change. Those funds can be used to replenish and sustain the commons, building a wellbeing economy from the revamped tools of the old extractive one.

But how do we incorporate that sustenance and benefit into the fabric of our new operating system? For a wellbeing economy to function well, citizens need to benefit from their commons. Given that our commons have been enclosed (read "stolen" by the few), we are surely entitled to seek compensation or a share of them. This dividend would come from those economic activities by right of our citizenship, not as a "welfare transfer."

In *Plunder of the Commons*, labor and social policy expert Guy Standing identifies three main types of commons that would attract levies.[10] The first: "exhaustible commons," the nonrenewables like minerals and fossil fuels that should be treated as common natural capital assets. The second: "replenishable commons" that require funds to be set aside for their replenishment. Last are the "renewable commons" that range from tangibles like water and the atmosphere to intangibles like ideas. Fees earned from all these commons could be made available for immediate distribution back to everyone. This would be compensation for enclosing the commons.

In addition, just as in the examples of Nepal and the Charter of the Forests, the commons should be able to contribute to subsistence, to meeting basic needs. Access or minor use under a commons banner—which is always about maintaining the value of the commons stock, or asset—is surely to be encouraged. The commons allow us to find a way of participating in economic activity to meet our needs and those of our community. This activity often slips off the radar as it is usually incredibly small-scale, very often a peer-to-peer exercise.

In Latin America, the Transnational Institute lists the criteria that underpins a wellbeing economy based on the commons: "(1) material or immaterial resources managed collectively and democratically, (2) social processes that foster and deepen cooperative relationships,

(3) a new logic of production and a new set of productive processes, and (4) a paradigm shift, which conceives the commons as an advance beyond the classical market/state or public/private binary oppositions."[11] That last point is a recurrent theme: neither market nor state, and perhaps the better for that.

Indeed, new commons can be established and managed with sufficiency and optimism in mind—whether through a social enterprise, a trust, a cooperative, or simply a user group. A host of initiatives, many crystallized by digital infrastructure, have emerged in recent years. Such diverse commons include seed-sharing cooperatives; communities of open-source software programmers; the creation and use of complementary currencies to stimulate local economies; and local food initiatives such as community-supported agriculture, rewilding, Slow Food, and community land trusts. They all reflect the value-creating aspect of having accessible resources and tools outside private or state governance. Michel Bauwens, founder of the P2P Foundation, focusing on peer-to-peer economics, also describes them as "schools for democracy"—real exercises in participation and collaboration.[12]

How to Resolve the Systems Failure

A rentier economy is all about gatekeeping access to stocks and the value of stocks to the benefit of owners. Taxation doesn't impact the overuse of land and minerals (through resource taxes) significantly either. An additional toolbox is needed to capture some of the economic rents as fees and direct them to become part of a citizens' dividend.

An example of such a dividend was developed by economist James Boyce and entrepreneur Peter Barnes to address carbon pollution and the use of our atmosphere. They propose charging a fee as far up the supply chain as possible. If the atmosphere is polluted by carbon emissions, then a cap on emissions (for example) should be associated with a sufficiently high price per metric ton to drive polluters to reduce their assault on our common resource of a stable climate. This would imply a rising price on carbon emissions. So far, so familiar. But at this stage, it is not really about the commons; it is simply a kind of tax on the "bads" (pollution, or, in economic jargon, an "externality"). Such taxes are clearly a good idea. However, the impact of much higher energy prices has a disproportionate impact on the poor, and those in Most of the World who did not cause a significant part of global warming. This has always reduced the appeal of resource taxes, and with very good reason.

If a fee and dividend approach is introduced, then the proceeds from the large increase in the price of carbon are recirculated to everyone, as the co-owners. It becomes a universal basic dividend. This helps compensate those on low incomes, who have extremely low carbon emissions compared with wealthy people, while damping down carbon-heavy activity by companies. It incentivizes conservation-oriented choices everywhere. To make this authentic and trustworthy, in a context when few trust either government or corporations, the fees and distribution could be handled by a Citizens Fund set up as a fiduciary trust alongside the central bank. Such a new institution would be dedicated to this one task, and if well managed should garner confidence. It may seem a bold step today, but central banks are already exploring digital currencies with accounts for everyone. And other semi-autonomous institutions already exist.

The carbon fee/dividend is an example of a process that taxes unearned income or economic rents, not earned incomes. It is about recognizing asset classes potentially as diverse as our personal data, land value uplift created primarily by location, financial infrastructure, and the backbone networks, like the Internet, that were created by governments but have been appropriated by private owners. It is a key tool for economic justice and has the potential to gain huge political support as a consequence.

Alaska's Permanent Fund is a genuinely universal fund, as every man, woman, and child is entitled to their share. It varies with the fees gathered for each year, and is sensitive to market realities. It is not a welfare payment; it's not supposed to do more than add to existing income and benefits. It is popular, and seen as a matter of economic justice and inclusion of all citizens.

The basic concept of fee/dividend was supported by 3,000 economists when it appeared in 2019 as a bipartisan proposal in the US called the Baker-Schultz plan. If it is less well-known than it should be, it is not without considerable support.

In his book *With Liberty and Dividends for All*, Barnes[13] suggests that a comprehensive fee and dividend system imposed predominantly on carbon, financial infrastructure transaction fees, and fees related to intellectual property has the potential to pay $5,000 a year to each American citizen. In the US, where the median annual income is approximately $80,000, a family of four could receive an additional $20,000 from a Citizens Fund.

This would be a significant addition to incomes during a period of disruptive transformation—while reducing overconsumption of resources like fossil energy in a fair way. It is clear that good policy design around Citizens Funds could be a game-changing economic innovation to build trust, goodwill, and economic security in a time of crisis.

It is scenarios like these that underscore that there is more to a commons perspective than dividends. What is required when thinking as co-owners (and trustees) of a commons is not complicated to understand. It has just three main strands:

- A chance to participate in the economy, with access to tools and resources, as well as to add value beyond just employee or customer.
- A share or dividend of the outcomes of any enclosure (as a co-owner is due a dividend).
- A commitment to ensuring that this dividend endures—through capital maintenance or enhancement of the resource that provides it.

Conclusions

It comes as no surprise that so many of the turnarounds discussed in this book have focused on readjusting flows within today's economic systems. The energy turnaround requires the combination of technologies, behavioral change, and price to accelerate change and attract investments through public and private channels. The same pattern underpins the food turnaround. Very rapid innovation cycles for regenerative agriculture, precision fermentation, and cellular agriculture will both attract investment and lower prices thus generating greater demand: a virtuous cycle. Three of the five turnarounds are about redistribution: namely the poverty, inequality, and empowerment turnarounds. Together, the three make the economy more inclusive and strengthen social capital.

We have also acknowledged that the economy we experience has increasingly become a monetary phenomenon. And while we propose a shift to a commons-based wellbeing economy, it is worth contemplating that many of the tools that enable this shift are rooted in our existing system. It is a delicate balance between realism and idealism, upgrading and transforming, evolution and revolution.

In addressing poverty, the role of the monetary economy is central: International institutions like the IMF and World Bank could create and use existing instruments—like Special Drawing Rights—and the cancellation of onerous debt to create a clean slate and protect the bedrock domestic development of low- and middle-income nations. This could be coupled with a reform of the trading rules. Both these shifts are embedded in the mores of the recent past.

Even the turnarounds focused on the empowerment of women and a reduction in inequality include expanding existing social programs, adjusting taxes, and appropriate legislation—the sorts of activities that have, over decades, made slow but still insufficient progress but could make significant change with more effort.

Yet all these changes lead to a more profound look at economic relationships and at the nature of economics itself. They are exactly what Donella Meadows[14] described: "places within a complex system…where a small shift in one thing can produce big changes in everything." Used well, and coupled with a renewed and reimagined focus on the commons, these tools and the radical shifts they enable can close the loop on an extractive economic system and make it not just circulatory but also regenerative. It can reduce our material footprints and help people steward the Earth that sustains them. And it can put the economy back in service to people.

We call it a transformational wellbeing economics. Earth for all.

9

A Call to Action

Dear readers, you came, you read. Thank you.

We hear you. The tasks are monumental. The barriers are huge. The risks are profound. The timeline is short. We have been talking about catalyzing the fastest economic transformation in history. And the heavy lifting has to start in the first decade. Right now. As you close this book.

It is probably worth reminding ourselves what we will gain from upgrading and rebooting our economic system.

An end to poverty in one generation
This is now within reach. We estimate that all countries could have average national incomes above $15,000 per person per year by 2050. Snoozing means waiting a generation later—approaching 2100 to reach this monumental goal.

Greater equality among people and nations
Our societies will not be torn apart by excessive inequalities. Redistributing wealth within and among nations means that future generations can have more chances to realize their dreams, irrespective of their family or country of origin.

Healthy people on a healthy planet
All people have the choice to eat well. Healthy food is the foundation of a good, long, healthy life. It is also the foundation of a livable planet. The alternative means reaching an uncomfortable tipping point this century when over half the planet will be overweight or obese, while hundreds of millions go hungry.

Abundant clean, cheap energy
By 2050, most countries could have abundant clean energy for the first time and at a price considerably cheaper than today's energy. And for the first time, most could have energy security. This will allow countries to break away from uncomfortable relationships with authoritarian regimes that control fossil fuel supplies.

Fresh air
The toxic brown clouds hanging over major cities will disappear. The shift to clean energy and energy efficiency brings with it a dramatic fall in air pollution. Without it, the markets will "solve" the problem of air pollution with "solutions" such as the bubble schools in our four girls' narratives (these schools exist today!)—where children of rich parents can play in playgrounds with filtered air, while other children breathe polluted air.

Gender equity
The dismantling of the patriarchal hierarchy will really help drive wellbeing for all and human development. Gender equity helps build social cohesion because diversity, fairness, and justice are valued.

Economic resilience and security
Upgrading the economic system so that all people more fairly share in the common wealth of societies will build resilience to inevitable shocks; improve trust in democracies, making them better equipped to take long-term decisions that benefit the majority of people; and build support for an almighty economic reboot.

Population stabilization
Within a single generation, the population of the planet can peak—potentially below nine billion—and begin to fall in the second half of the century. Succeeding here, in a way that is fair and just and supported by people—that is, through providing economic security and promoting gender equity—will be one of the most important achievements in the history of humanity.

A livable planet
Our future will be vastly more peaceful, more prosperous, and more secure if we do everything in our power to stabilize Earth by 2050, starting today. The longer we wait, the more dangerous our future becomes. Our proposal is intended to help ensure a future on a livable planet, a relatively stable planet with resilient societies better equipped to adapt to change. If anything, this book is a response to their call for "systems change".

We gain back our future
Most of all, what we gain is our future. The foundations of a vibrant economy are not money, nor energy, nor trade; they are optimistic people with hope for a better future...and the tools to create this future.

Is Earth for All Closer Than We Think?

If you feel the scale of the transformation is daunting, join the club. Perhaps you feel it is like pushing a boulder up a hill. Well, here we have some good news. You will have to push a boulder, for sure, but what if you have to push the boulder downhill instead of uphill? What if we just need to get the damn thing moving and the force of gravity will help us after that?

We believe we are reaching a social tipping point across societies. Indeed, four forces—social movements, new economic logic, technological development, and political action—are already aligning to push societies across a tipping point in a way that leads us to self-reinforcing virtuous cycles, an Earth for All world.

Social movements—the voice of the future
In 2018, Greta Thunberg began her strike from school outside the Swedish parliament building. Her protests about inaction on climate were joined by other young people around the world. From nowhere, the future had a voice and it was loud and angry. Around the same time, other movements have emerged or grown to prominence—#MeToo, Black Lives Matter, Sunrise, and Extinction Rebellion among

them. Public awareness has never been so high. These inspirational movements are shaping the public debate and forcing politicians to sit up and seriously take a systems approach to existential risk, for the first time.

Crossing an economic tipping point

For too long, those arguing against systems change have said it is too expensive. In recent years, the logic behind this argument has disappeared. In many places, it is now cheaper to build a solar array than it is to keep a coal-fired power station running. And getting cheaper every year. The cost of wind power is dropping precipitously too. The cheapest sources of electricity in history are now renewables. Even without strong economic policies to promote clean technology, the transition is inevitable and will happen faster than many predictions just a few years ago. But we can't afford delays.

Technology—disruption is coming

The Fourth Industrial Revolution is underway and will accelerate this decade. Digitalization and other technologies including automation, artificial intelligence, and machine learning will have a disruptive impact across all industries, which will change demand for products, change the nature of work, and change societies in ways that are difficult to predict. Harnessed and directed, the technological revolution can dramatically drive down demand for energy; help provide sustainable food; improve gender equity through changing how we work and live our lives; and it can reduce poverty by connecting more people into the global economy.

Accelerating political momentum

Politicians have been shaken awake to a new narrative driven by youth movement, new economic logic, and technological breakthroughs. Most major economies have now committed to reach net-zero emissions by 2050 (and 2060 and 2070 in the cases of China and India). Some countries are embracing wellbeing economies: Finland, Iceland, New Zealand, Scotland, and Wales. Europe's Green Deal promises a fair and just transition to a zero-carbon future. Countries such as Spain are investing in protection of coal workers to help them retrain during the transition period. In the United States, the Green New Deal, again built on the foundation of a fair and just transition, is gaining momentum. And China's Ecological Civilization is a profound long-term economic narrative based on a society operating in harmony with nature. The time is right for extraordinary turnarounds.

Given these social tipping points, the boulder we need to push may only require a large shove to really get it moving, with its own unstoppable momentum. And when we push the boulder, we may have a lot of hands pushing with us. At the start of the Earth for All initiative, we commissioned market research company Ipsos MORI to run an extensive international survey of G20 countries.[1] Ipsos MORI recruited around 20,000 people across these countries to complete the survey. The findings are informative and even provide a beacon of hope. The world is not sleepwalking toward catastrophe. People are well aware of the colossal risks we are taking by continuing business as usual.

- Across the G20, three in five people (58%) are "extremely worried" or "very worried" about the current state of the planet. Even more are concerned about the future. Concern is at its highest level among: women (62%), young people aged 25 to 34 (60%), those educated, high earners, and people who identify more as global citizens than those who have a very strong national identity.

- Three in four people (73%) believe Earth is edging closer to tipping points because of human action. Those living in places close to large, vital ecosystems currently under attack from development, like Indonesia and Brazil's rainforests, are most aware.

- Do people want to become better planetary stewards? Are people willing to do more to protect nature and the climate? Again, the answer is a resounding, "Yes" (83%). So most people across the world's largest economies do indeed want to do more to protect and restore nature. But that does not mean that they are willing to pay the bill.

The biggest surprise in the survey came in response to our asking people if they want to transform economic systems to prioritize well-being, health, and protection of the planet over a singular focus on profit and economic growth. The answer is once again a resounding "Yes." Among G20 countries, 74% of people support the idea that their country's economic priorities should move beyond profit and increasing wealth and focus more on human wellbeing and ecological protection. This view is consistently high among all G20 countries. It is particularly high in Indonesia (86%), and even in the lowest-scoring countries like the United States (68%), people support change.

A Chorus of Voices

The solutions we have described in this book require big changes in societies everywhere. Those of us who have responsibilities in governments and international institutions, or the private and financial sector, are already in privileged positions to become champions of this transformation. Ultimately, though, the solutions are largely macroeconomic and so require governments to create new policies to make stuff happen. Things like progressive taxation, creating a Citizens Fund, redesigning the International Monetary Fund, or transforming the energy system are beyond the power of individuals or even banks or major companies to deliver at the scale needed. (See Fifteen Policy Recommendations sidebar.)

Fifteen Policy Recommendations

Poverty
- Allow the International Monetary Fund to allocate over $1 trillion annually to low-income countries for green jobs—creating investments through so-called Special Drawing Rights.
- Cancel all debt to low-income countries (<$10,000 income per person).
- Protect fledgling industries in low-income countries and

promote South-South trade between these countries.
Improve access to renewables and health technologies
by removing obstacles to technology transfer, including
intellectual property constraints.

Inequality

- Increase taxes on the 10% richest in societies until they take
  less than 40% of national incomes. The world needs strong
  progressive taxation; and closing international loopholes
  is essential to deal with destabilizing inequality and luxury
  carbon and biosphere consumption.
- Legislate to strengthen worker's rights. In a time of deep
  transformation, workers need economic protection.
- Introduce Citizens Funds to give all citizens their fair share
  of the national income, wealth, and the global commons
  through fee and dividend schemes.

Gender Equity

- Provide access to education for all girls and women.
- Achieve gender equity in jobs and leadership.
- Provide adequate pensions.

Food

- Legislate to reduce food loss and waste.
- Scale up economic incentives for regenerative agriculture
  and sustainable intensification.
- Promote healthy diets that respect planetary boundaries.

Energy

- Immediately phase out fossil fuels and scale up energy
  efficiency and renewables. Triple investments immediately
  to >$1 trillion per year in new renewables.
- Electrify everything.
- Invest in energy storage at scale.

We want to close with a call to action. Sometimes change can seem impossibly slow, generational even. But it does not need to be so. The global financial crisis that played out between 2007 and 2009 drove extremely rapid political and economic change to build more resilient banking systems. The COVID-19 pandemic led to overnight changes in behavior and business models. This gives us hope that this decade will see the fastest economic transformation in history.

All of us have a role to play, as concerned citizens, as human beings, as people who value our future, to support this change. Politicians respond to public voices, and the pathways we are advocating will need public action and a chorus of voices to reach unstoppable momentum. We need a movement of movements built on outrage and optimism. We need a change in the narrative, we need to open a conversation in every home, every school, every university, every town and city on how to upgrade our economic system. We think this is possible. At the end of the day, it is about defending our common sacred values, providing a home for our families, our kids, our loved ones, ensuring the dignity of each human being, and looking forward to the future on a livable planet.

The call to action for governments is this: commit to the five extraordinary turnarounds and the accompanying policy levers. To create momentum to drive these turnarounds:

- Reduce polarization. Improve social cohesion. Find common ground. Or lose democracy.
- Share wealth more fairly. Citizens Funds and Universal Basic Dividends bring multiple benefits. They are likely to gain the support from the majority of people. They reduce harmful pollution. And they protect citizens during a turbulent period.
- Act in the interests of future generations and create institutions that ensure current generations think intergenerationally.
- Change how you measure progress, valuing wellbeing over financial growth.
- Engage with citizens about what really matters in society.
- Send unequivocal signals to markets that long-term commitment and investment in transformation is locked in. This will create economic optimism for transformation.

The call to action for citizens is this:

- Join the movements!
- Vote for politicians who value the future.
- Wherever you are, start conversations about how the economic transformation that has to come will affect you, your family, your job, your life. How can you benefit from it? How can it improve your career, your education? Is this societal transformation an opportunity to follow your dream, to change course?
- In your town, or city, or country, demand a citizens' assembly on economic systems change. Citizens' assemblies have been used to navigate difficult, contentious political issues like climate change. They can help diffuse polarization and provide fresh ideas and perspectives. We believe they are one of the most exciting ways to get politicians to sit up and take notice.
- Ask your local and national politicians to act to bring our societies closer to Earth for All.

Across our tribes, cultures, and societies, people everywhere are worried and anxious about the future. But we have two things in common: We all value our future, and the majority of people want change and want support to change. If this book does one thing, we hope to convince you there is a future out there worth striving toward. This future is not some bright utopia without shadows. But we believe the pathways discussed here are the ones most likely to secure the long-term potential of our wondrous, freewheeling, endlessly inventive, often confounding, and truly global civilization on a relatively stable planet.

Appendix

The Earth4All Model

As everyone who works methodologically and professionally in foresight knows, the future does not exist yet, so there can be no evidence-based data from the future (until we get there).

So scenarios are narratives about the future. They are plausible and uncertain stories. And numbers from the future are metaphors in that story, not absolute truths about some predetermined reality. Those numbers are sound bites about the future intended to inform the decisions we make in the present.

This applies, of course, to all models that calculate, assess, or estimate climate change, demographics, or anything else into the distant future.

Here, we share some of the workings of Earth4All that helped us visualize Too Little Too Late, Giant Leap, and other scenarios that were part of our project analysis. And we invite you to access the data and use it for your own investigations.

Model Purpose

Earth4All is a system dynamics computer model to study the dynamics of human wellbeing on finite planet Earth this century. The model is designed primarily to generate internally consistent scenarios for population, poverty, GDP, inequality, food, energy, and other relevant variables from 1980 to 2100 and to see how they evolve in concert. The ambition is to identify policies that increase the likelihood of a future that combines high wellbeing for the global majority with thriving nature keeping Earth within planetary boundaries.

Two main versions of the model exist: One calculates global averages (E4A-global), and the other calculates development paths for ten regions of the world (E4A-regional).

The model was built to recreate the broad sweep of history from 1980 to 2020, using as few exogenous drivers as possible. An exogenous driver is where the value of a variable in the model is imposed externally—we turn up a knob on the model manually—rather than generated internally by the dynamics of the model, for instance, if a nation or region increases taxes on the top earners in society, the model responds by calculating how this driver impacts the financial system and available income to the government into the future.

We can run the model to 2100 to study the effect of parameter changes made in 2022. The model has intentionally been kept as simple as possible, to increase transparency and understandability, even though this lowers the precision level.

Model History

The Earth4All model has been under construction for a decade. It was (of course) inspired by the World3 model that underlaid *The Limits to Growth* and, in different versions, the follow-up studies in 1992 and 2004. In 2011, two of us (Jorgen Randers and Ulrich Goluke) began a concerted effort to develop a system dynamics model to include investment needs that might help actually solve some of the existential challenges World3 highlighted so effectively, specifically the unavoidable nonprofitable spending required to solve the climate emergency. We did not succeed in turning our work into a system dynamics model, but we successfully created a regionalized spreadsheet model, called Earth2, which supported the book *2052: A Global Forecast for the Next Forty Years*, which appeared in 2012.[1] During the following years, Earth2 was gradually improved into the Earth3 model, which supported the book *Transformation Is Feasible!* published in 2018.[2] In parallel, we evolved the climate component of Earth2 into a fully-fledged system dynamics model of climate change this century, published in 2016.[3]

The (big) job of converting these earlier models into a fully endogenized system dynamics model was completed during the Earth for All project over the last several years and given the name the Earth4All model.

The Main Sectors in the Model

The model consists of the following sectors (in the regionalized model, one for each of the regions):

- **Population sector:** generates total population from fertility and mortality processes, potential workforce size, and the number of pensioners.
- **Output sector:** generates GDP, consumption, investment, government spending, and jobs. The economy is seen as a sum of a private sector and a public sector.
- **Public sector:** generates public spending from tax revenue, the net effect of debt transactions, and the distribution of the budget on governmental goods and services (including on technological advance and the five turnarounds).
- **Labor market sector:** generates the unemployment rate worker share of output, and the workforce participation rate, based on the capital output ratio.
- **Demand sector:** generates income distribution between owners, workers, and the public sector.
- **Inventory sector:** generates capacity utilization and the inflation rate.
- **Finance sector:** generates the interest rates.
- **Energy sector:** generates fossil fuel-based and renewable energy production, greenhouse gas emissions from fossil fuel use, and the cost of energy.
- **Food and land sector:** generates crop production, environmental impacts of agriculture, and the cost of food.
- **Reform delay sector:** generates the societal ability to react to a challenge (like climate change) as a function of social trust and social tension.

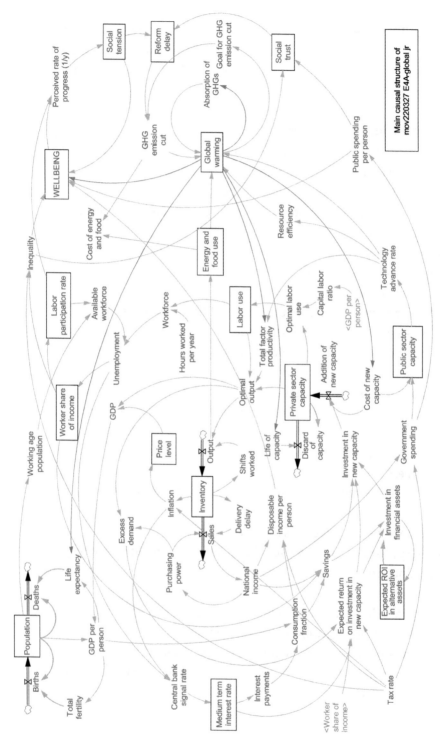

Figure A.1 The main causal loops that make up the core structure of the Earth4All model.

- **Wellbeing sector:** generates global indicators measuring both environmental and societal sustainability. Including the Average Wellbeing Index.

In the regionalized version of Earth4All, world development is calculated as the sum of the development in ten regions (each being represented by the same structure and parametrized to fit their regional style of socioeconomic development). When fitting to the regions, we learned that numerous Sustainable Development Goals and behavior traits vary systematically as a function of GDP per person.[4] This applies to savings rate, fertility, life expectancy, pension age, energy use per person, food use per person, minerals use per person, hours worked per year, and other factors.

Model Causal Loop Diagram
A high-level description of the model structure is shown in figure A.1.

A full technical description of the model is available on the web at earth4all.life. This includes the model's underlying equations. Users can also download and run the open-source model on their own computers.

Model Novelty
Why make a new model when there are so many others out there already? What unique attributes does the Earth4All model offer? Here we list eight novelties that address some of the shortcomings in the global systems modeling field:

1. **Inequality:** We investigate the distributional effects in terms of owner and worker share of output from both private investment and public sector activities, confirming the preliminary evidence that distributional patterns are relevant to sustainable policy-making.[5]
2. **Ecology:** We include the wider effect of the human economy on the main planetary boundaries (climate, nutrients, forests, biodiversity), the impact of natural boundaries on economic development, and their complex feedback effects.[6]

3. **Public sector:** We model an active public sector with public infrastructure capacity, welfare policies, and climate-change mitigation policy stance.[7]

4. **Finance:** We include the effects from debt and money supply, central bank interest rates, and corporate capital costs, addressing the call for further integration of financial mechanisms within integrated assessment models (IAMs), used to test the feasibility of climate goals.[8]

5. **Labor:** We are able to simulate a recurrent ten-year unemployment cycle, and its macroeconomic consequences, a global first.[9]

6. **Population:** In contrast to the UN's statistical approach, the Earth4All model has endogenous population dynamics affected by investment levels in public spending, education, and income levels, improving on existing IAMs with demographic sectors.[10]

7. **Wellbeing:** We integrate an Average Wellbeing Index (as a function of disposable income, income inequality, government services, the climate crisis, and perceived progress), illustrating the connection between environmental sustainability and social trust, and linking declining trust to public decision-making delays in an integrated assessment model for the first time.[11]

8. **Social tension:** We integrate a Social Tension Index (as a function of perceived progress, defined as the rate of change in the Average Wellbeing Index) that influences the speed and strength at which societies react to an emerging challenge. That is, as the Social Tension Index rises, we interpret this as driving greater polarization in societies, making it more challenging to agree on solutions to societal challenges like the climate emergency.

The Earth for All Game

The Earth4All model will also become freely available with a user-friendly, intuitive interface. This is to make it simpler for those who want to run the model with their own set of parameters to do so.

Playing with models is a helpful way to learn how the system's dynamics work, and how certain changes stimulate other changes. In interactive sessions, classes, groups, and citizen assemblies can negotiate and game together to create their own future.

Notes

Chapter 1: Earth for All

1. Sub-Saharan Africa, South Asia, Southeast Asia, China, western Europe, eastern Europe and central Asia, Latin America, Middle East and North Africa, Pacific region, and United States.
2. By 2100, the Too Little Too Late scenario sees an approximate 2.5°C rise in global average surface temperature above preindustrial levels.
3. This is the rough assessment that arises from all the assumptions that constitute the Earth4All model. It is supported by other studies on the likely cost of action. See, for example: the International Energy Agency's *Net Zero by 2050: A Roadmap for the Global Energy Sector* (2021); the Intergovernmental Panel on Climate Change's "Mitigation Pathways Compatible with 1.5°C in the Context of Sustainable Development," Chapter 2 in: *Global Warming of 1.5°C. An IPCC Special Report* (2018); Yuval Noah Harari, "The Surprisingly Low Price Tag on Preventing Climate Disaster," *Time* (January 18, 2022); DNV's *Energy Transition Outlook–2021* (Oslo: DNV, 2021); Nicholas Stern's "Economic Development, Climate and Values: Making Policy," *Proceedings of the Royal Society* 282, no. 1812 (August 7, 2015).
4. Chandran Nair, *The Sustainable State: The Future of Government, Economy, and Society* (Oakland, CA: BK Publishers, 2018); Mariana Mazzucato, *Value of Everything* (S.l.: Public Affairs, 2020).
5. Donella H. Meadows et al., *The Limits to Growth: A Report for the Club of Rome's Project on the Predicament of Mankind* (New York: Universe Books, 1972). This report was commissioned by the Club of Rome. Despite its name, this is an international think tank dedicated to systemic systems thinking about global problems.
6. Graham M. Turner, "On the Cusp of Global Collapse? Updated Comparison of *The Limits to Growth* with Historical Data," *GAIA-Ecological Perspectives for Science and Society* 21, no. 2 (2012): 116–24; Graham Turner, *Is Global Collapse Imminent?* MSSI Research Paper No. 4, (Melbourne Sustainable Society Institute, University of Melbourne, 2014).
7. Gaya Herrington, "Update to Limits to Growth: Comparing the World3 Model with Empirical Data," *Journal of Industrial Ecology* 25, no. 3 (June 2021): 614–26.
8. Colin N. Waters et al., "The Anthropocene Is Functionally and Stratigraphically Distinct from the Holocene," *Science* 351, no. 6269 (2016).
9. Paul J. Crutzen, "Geology of Mankind," *Nature* 415, no. 6867 (2002): 23.
10. A. Ganopolski, R. Winkelmann, and H. J. Schellnhuber, "Critical Insolation–CO_2 Relation for Diagnosing Past and Future Glacial Inception," *Nature* 529, no. 7585 (2016): 200–203.

11. Will Steffen et al., "The Trajectory of the Anthropocene: The Great Acceleration," *Anthropocene Review* 2, no. 1 (April 2015): 81–98.
12. Will Steffen et al., "Planetary Boundaries: Guiding Human Development on a Changing Planet," *Science* 347, no. 6223 (2015).
13. Lan Wang-Erlandsson et al., "A Planetary Boundary for Green Water," *Nature Reviews Earth & Environment* (2022): 1–13.
14. Timothy M. Lenton et al., "Climate Tipping Points: Too Risky to Bet Against," *Nature* 575 (2019): 592–95; Jorgen Randers and Ulrich Goluke, "An Earth System Model Shows Self-Sustained Thawing of Permafrost Even If All Man-Made GHG Emissions Stop in 2020," *Scientific Reports* 10, no. 1 (2020): 18456.
15. Kate Raworth, *Doughnut Economics: Seven Ways to Think Like a 21st-Century Economist* (VT: Chelsea Green, 2017).
16. In Earth4All, when we write $, USD, or US$, the default we refer to is US dollars at stable, real 2017 prices, measured at purchasing power parity (PPP)—in this case, 15,000 US$ 2017-PPP—building on the Penn World Tables v.10.
17. See Dr. Mamphela Ramphele (2022), Deep Dive paper, *Global Equity for a Healthy Planet*, available at earth4all.life/resources.
18. Emily Elhacham et al., "Global Human-Made Mass Exceeds All Living Biomass," *Nature* 588, no. 7838 (December 2020): 442–44.
19. Paul Fennell et al., "Cement and Steel: Nine Steps to Net Zero," *Nature* 603, no. 7902 (March 2022): 574–77.
20. Ibid.
21. Argentina, Australia, Brazil, Canada, China, France, Germany, Great Britain, India, Indonesia, Italy, Japan, Mexico, Russia, Saudi Arabia, South Africa, South Korea, Turkey, the United States.

Chapter 2: Exploring Two Scenarios

1. See Randers et al. (2022), *The Earth4All Scenarios* Technical report, earth4all.life/resources.
2. For more information, see weall.org.
3. David Collste et al., "Human Well-being in the Anthropocene: Limits to Growth," *Global Sustainability* 4 (2021): e30.
4. Manfred A. Max-Neef, *Human Scale Development: Conception, Application and Further Reflections* (NY: Apex, 1991); Len Doyal and Ian Gough, "A Theory of Human Needs," *Critical Social Policy* 4, no. 10 (1984): 6–38.
5. See Richard Wilkinson and Kate Pickett (2022), Deep Dive paper, *From Inequality to Sustainability*, available at earth4all.life/resources.
6. Jon Reiersen, "Inequality and Trust Dynamics," in *Disaster, Diversity and Emergency Preparation*, ed. Leif Inge Magnussen (NATO/IOS Press, 2019).
7. See L. Chancel et al., *World Inequality Report 2022* (World Inequality Lab, 2021).
8. Eric Lonergan and Mark Blyth, *Angrynomics* (Newcastle upon Tyne, UK: Agenda, 2020).
9. The Earth4All project's Too Little Too Late scenario has a climate trajectory that is close to the "Middle of the road" scenario in the Shared Socioeconomic Pathway family of scenarios, i.e., the "IPCC SSP2-4.5" scenario.

See Malte Meinshausen et al., "The Shared Socio-Economic Pathway (SSP) Greenhouse Gas Concentrations and Their Extensions to 2500," *Geoscientific Model Development* 13, no. 8 (August 13, 2020): 3571–3605.

10. For more on fee and dividends, see chapter 4 on the inequality turn-around, chapter 8 on wellbeing economics, and the full Deep Dive paper by Ken Webster (2022), *The Long Road to a Social Dividend*, available at earth4all.life/resources.

11. Ngũgĩ wa Thiong'o, *Decolonizing the Mind: The Politics of Language in African Literature* (London, Nairobi: J. Currey Heinemann Kenya [etc.], 1986).

Chapter 3: Saying Goodbye to Poverty

1. IRP et al., *Global Resources Outlook 2019: Natural Resources for the Future We Want* (UNEP/IRP, 2019).

2. B. Bruckner et al., "Impacts of Poverty Alleviation on National and Global Carbon Emissions," *Nature Sustainability* 5 (April 2022): 311–20.

3. Henry A. Giroux, "Reading Hurricane Katrina: Race, Class, and the Biopolitics of Disposability," *College Literature* 33, no. 3 (2006): 171–96.

4. Nishant Yonzan, Christoph Lakner, and Daniel Gerszon Mahler, "Projecting Global Extreme Poverty up to 2030," *World Bank Blog* (October 9, 2020).

5. Masse Lô (2022), *Growth Within Limits Through Solidarity and Equity*, Earth for All Deep Dive paper, available at earth4all.life/resources.

6. K. Sahoo and N. Sethi, "Impact of Foreign Capital on Economic Development in India: An Econometric Investigation," *Global Business Review* 18, no. 3 (2017): 766–80; S. Sharma et al., "A Study of Relationship and Impact of Foreign Direct Investment on Economic Growth Rate of India," *International Journal of Economics and Financial Issues* 10, no. 5 (2020): 327; A. T. Bui, C. V. Nguyen, and T. P. Pham, "Impact of Foreign Investment on Household Welfare: Evidence from Vietnam," *Journal of Asian Economics* 64 (October 2019): 101130.

7. J. Zheng and P. Sheng, "The Impact of Foreign Direct Investment (FDI) on the Environment: Market Perspectives and Evidence from China," *Economies* 5, no. 1 (March 2017): 8.

8. World Bank (2022), "International Debt Statistics | Data."

9. Paul Brenton and Vicky Chemutai, *The Trade and Climate Change Nexus: The Urgency and Opportunities for Developing Countries* (Washington, DC: World Bank, 2021).

10. "Lawrence Summers' Principle," ejolt.org/2013/02/lawrence-summers'-principle.

11. See Jayati Ghosh et al. (2022), Deep Dive paper, *Assigning Responsibility for Climate Change: An Assessment Based on Recent Trends*, written with co-authors from Political Economy Research Institute, University of Massachusetts Amherst, US, available at earth4all.life/resources.

12. Jayati Ghosh "Free the Money We Need," *Project Syndicate* (February 14, 2022).

13. An important recent case of this can be seen in the efforts by drug companies to discourage bringing COVID-19 vaccines to Africa, see Madlen Davies, "COVID-19: WHO Efforts to Bring Vaccine Manufacturing to

Africa Are Undermined by the Drug Industry, Documents Show," *BMJ* 376 (2022): 0304.

14. Anuragh Balajee, Shekhar Tomar, Gautham Udupa, "COVID-19, Fiscal Stimulus, and Credit Ratings," *SSRN Electronic Journal* (2020).

15. "Home–Commission of Inquiry into Allegations of State Capture," accessed April 7, 2022, statecapture.org.za.

Chapter 4: The Inequality Turnaround

1. In addition to taxation, there is an urgent need for regulation of markets and investor behavior to align private investments with social goals, prevent excessive concentration, and reduce monopolistic behavior and rent seeking by large corporations.

2. L. Chancel et al., *World Inequality Report 2022* (World Inequality Lab, 2021).

3. Michael W. Doyle and Joseph E. Stiglitz, "Eliminating Extreme Inequality: A Sustainable Development Goal, 2015–2030," *Ethics & International Affairs* 28, no. 1 (2014): 5–13.

4. See Wilkinson and Pickett (2022), Deep Dive paper, *From Inequality to Sustainability*, available at earth4all.life/resources.

5. Chancel et al., *World Inequality Report 2022*.

6. Wilkinson and Pickett (2022).

7. Chancel et al., *World Inequality Report 2022*.

8. Oxfam (September 21, 2020), Media Briefing, "Confronting Carbon Inequality: Putting Climate Justice at the Heart of the COVID-19 Recovery."

9. See Chandran Nair (2022), Deep Dive paper, *Transformations for a Disparate and More Equitable World*, available at earth4all.life/resources.

10. Alex Cobham and Andy Sumner, "Is It All About the Tails? The Palma Measure of Income Inequality," Center for Global Development Working Paper No. 343, *SSRN Electronic Journal* (2013).

11. Chris Isidore, "Buffett Says He's Still Paying Lower Tax Rate Than His Secretary," *CNN Money* (March 4, 2013).

12. Lawrence Mishel and Jori Kandra, *CEO Pay Has Skyrocketed 1,322% Since 1978* (Economic Policy Institute, August 10, 2021).

13. Climate Leadership Council, "The Four Pillars of Our Carbon Dividends Plan," accessed March 31, 2022; "Opinion | Larry Summers: Why We Should All Embrace a Fantastic Republican Proposal to Save the Planet," *Washington Post* (February 9, 2017), accessed March 31, 2022.

Chapter 5: The Empowerment Turnaround

1. Mariana Mazzucato, "What If Our Economy Valued What Matters?" Project Syndicate (March 8, 2022).

2. L. Chancel et al., *World Inequality Report 2022* (World Inequality Lab, 2021).

3. Max Roser, "Future Population Growth," Our World in Data (2022).

4. Wolfgang Lutz et al., *Demographic and Human Capital Scenarios for the 21st Century: 2018 Assessment for 201 Countries* (Publications Office of the European Union, 2018); see also Callegari et al. (2022), *The Earth4All Population Report to GCF*, at earth4all.life/resources.

5. Jumaine Gahungu, Mariam Vahdaninia, and Pramod R. Regmi, "The

Unmet Needs for Modern Family Planning Methods among Postpartum Women in Sub-Saharan Africa: A Systematic Review of the Literature," *Reproductive Health* 18, no. 1 (February 10, 2021): 35.

6. UNESCO, *New Methodology Shows 258 Million Children, Adolescents and Youth Are Out of School*, Fact Sheet no. 56 (September 2019).

7. Ruchir Agarwal, "Pandemic Scars May Be Twice as Deep for Students in Developing Countries," *IMFBlog* (February 3, 2022).

8. See Dr. Mamphela Ramphele (2022), Deep Dive paper, *Global Equity for a Healthy Planet*, available at earth4all.life/resources.

9. Sarath Davala et al., *Basic Income: A Transformative Policy for India* (London; New Delhi: Bloomsbury, 2015).

10. Andy Haines and Howard Frumkin, *Planetary Health: Safeguarding Human Health and the Environment in the Anthropocene* (NY: Cambridge University Press, 2021).

Chapter 6: The Food Turnaround

1. Cheikh Mbow et al., "Food Security," in *Climate Change and Land*, IPCC Special Report, ed. P. R. Shukla et al. (IPCC, 2019).

2. "Hunger and Undernourishment" and "Obesity," Ourworldindata.org, accessed February 20, 2022.

3. Yinon M. Bar-On, Rob Phillips, and Ron Milo, "The Biomass Distribution on Earth," *Proceedings of the National Academy of Sciences* 115, no. 25 (June 19, 2018): 6506–11.

4. Hannah Ritchie and Max Roser, "Land Use," *Our World in Data* (November 13, 2013).

5. M. Nyström et al., "Anatomy and Resilience of the Global Production Ecosystem," *Nature* 575, no. 7781 (November 2019): 98–108.

6. Bar-On et al., "The Biomass Distribution on Earth."

7. *The Future of Food and Agriculture: Alternative Pathways to 2050* (Food and Agriculture Organization of the United Nations, 2018), accessed March 31, 2022; *Climate Change and Land* (IPCC Special Report), accessed March 31, 2022.

8. Joe Weinberg and Ryan Bakker, "Let Them Eat Cake: Food Prices, Domestic Policy and Social Unrest," *Conflict Management and Peace Science* 32, no. 3 (2015): 309–26.

9. Rabah Arezki and Markus Brückner, *Food Prices and Political Instability*, CESifo Working Paper Series (CESifo, August 2011).

10. Lovins, L. Hunter, Stewart Wallis, Anders Wijkman, and John Fullerton. *A Finer Future: Creating an Economy in Service to Life*. Gabriola Island, BC, Canada: New Society Publishers, 2018.

11. Gabe Brown, *Dirt to Soil: One Family's Journey into Regenerative Agriculture* (VT: Chelsea Green, 2018).

12. Mark A. Bradford et al., "Soil Carbon Science for Policy and Practice," *Nature Sustainability* 2, no. 12 (December 2019): 1070–72.

13. T. Vijay Kumar and Didi Pershouse, "The Remarkable Success of India's Natural Farming Movement," Forum Network lecture (January 21, 2021).

14. https://www.rural21.com/fileadmin/downloads/2019/en-04/rural2019_04-S30-31.pdf

15. Johan Rockström et al., "Sustainable Intensification of Agriculture for Human Prosperity and Global Sustainability," *Ambio* 46, no. 1 (February 2017): 4–17.

16. Jules Pretty and Zareen Pervez Bharucha, "Sustainable Intensification in Agricultural Systems," *Annals of Botany* 114, no. 8 (December 1, 2014): 1571–96.

17. Andy Haines and Howard Frumkin, *Planetary Health: Safeguarding Human Health and the Environment in the Anthropocene* (NY: Cambridge University Press, 2021).

18. "Blue Food," *Nature*, nature.com (2021).

19. *The Future of Food and Agriculture*.

20. Sara Tanigawa, "Fact Sheet | Biogas: Converting Waste to Energy," EESI White Papers (October 3, 2017), accessed April 7, 2022.

21. Regulations should at least require implementation of the OECD/ILO Human Rights Due Diligence (HRDD) process.

22. Jules Pretty et al., "Global Assessment of Agricultural System Redesign for Sustainable Intensification," *Nature Sustainability* 1, no. 8 (August 1, 2018): 441–46.

23. Dieter Gerten et al., "Feeding Ten Billion People Is Possible Within Four Terrestrial Planetary Boundaries," *Nature Sustainability* 3 (January 20, 2020): 200–08.

24. Walter Willett et al., "Food in the Anthropocene: The EAT–Lancet Commission on Healthy Diets from Sustainable Food Systems," *Lancet* 393, no. 10170 (2019): 447–92.

Chapter 7: The Energy Turnaround

1. The Carbon Law is an exponential trajectory (halving every decade). Its name derives from another famous exponential trajectory, Moore's Law, in the digital technology sector, that observes computer power approximately doubles every two years.

2. Arnulf Grubler et al., "A Low Energy Demand Scenario for Meeting the 1.5°C Target and Sustainable Development Goals without Negative Emission Technologies," *Nature Energy* 3, no. 6 (June 2018): 515–27.

3. Jason Hickel, "Quantifying National Responsibility for Climate Breakdown: An Equality-Based Attribution Approach for Carbon Dioxide Emissions in Excess of the Planetary Boundary," *The Lancet Planetary Health* 4, no. 9 (September 1, 2020): e399–404.

4. Pierre Friedlingstein et al., *Global Carbon Budget 2021*, Earth System Science Data (November 4, 2021): 1–191.

5. See Jayati Ghosh et al. (2022), Deep Dive paper, *Assigning Responsibility for Climate Change: An Assessment Based on Recent Trends*, written with co-authors from Political Economy Research Institute, University of Massachusetts Amherst, US, available at earth4all.life/resources.

6. Benjamin Goldstein, Tony G. Reames, and Joshua P. Newell, "Racial Inequity in Household Energy Efficiency and Carbon Emissions in the United States: An Emissions Paradox," *Energy Research & Social Science* 84 (February 1, 2022): 102365.

7. Nate Vernon, Ian Parry, and Simon Black, *Still Not Getting Energy Prices*

Right: A Global and Country Update of Fossil Fuel Subsidies, IMF Working Papers (September 2021).

8. Grubler et al., "A Low Energy Demand Scenario."

9. See more in Janez Potočnik and Anders Wijkman (2022), Deep Dive paper, *Why Resource Efficiency of Provisioning Systems Is a Crucial Pathway to Ensuring Wellbeing Within Planetary Boundaries*, available at earth4all.life/resources.

10. Bill McKibben, "Build Nothing New That Ultimately Leads to a Flame," *New Yorker* (February 10, 2021).

11. Johan Falk et al., "Exponential Roadmap: Scaling 36 Solutions to Halve Emissions by 2030, version 1.5" (Sweden: Future Earth, January 2020).

12. Such commentators on the exponential rise of renewables include researcher groups at Stanford, IEA's Net Zero by 2050, IIASA; EWG & LUT, RMI, RethinkX, Singularity, Rystad, Statnett, Exponential View. See Nafeez Ahmed (2022), Deep Dive paper, *The Clean Energy Transformation*, available at earth4all.life/resources.

13. See Ahmed, Deep Dive paper, ibid., for discussion.

14. Ibid.

15. World Bank Group, *State and Trends of Carbon Pricing 2019* (Washington, DC: World Bank, June 2019): 9–10.

16. See Rimel I. Mehleb, Giorgos Kallis, and Christos Zografos, "A Discourse Analysis of Yellow-Vest Resistance against Carbon Taxes," *Environmental Innovation and Societal Transitions* 40 (September 2021): 382–94.

17. "Economists' Statement on Carbon Dividends Organized by the Climate Leadership Council," Original publication in the *Wall Street Journal*, econstatement.org.

18. Ghosh et al., 2022.

Chapter 8: From "Winner Take All" Capitalism to Earth4All Economies

1. Donella H. Meadows, *Leverage Points: Places to Intervene in a System* (Hartland, VT: Sustainability Institute, 1999).

2. Yuval Noah Harari, "The Surprisingly Low Price Tag on Preventing Climate Disaster," *Time* (January 18, 2022).

3. See Dr. Mamphela Ramphele (2021), Deep Dive paper, *Global Equity for a Healthy Planet*, available at earth4all.life/resources.

4. Elinor Ostrom, *Governing the Commons: The Evolution of Institutions for Collective Action*, Canto Classics (Cambridge, UK: Cambridge University Press, 2015); Peter Barnes, *Capitalism 3.0: A Guide to Reclaiming the Commons* (San Francisco: Berrett-Kohler, 2014).

5. Club of Rome, China Chapter (2022), "Understanding China" (forthcoming).

6. World Bank, *International Debt Statistics 2022* (Washington, DC: World Bank, 2021).

7. Per Espen Stoknes, *Tomorrow's Economy: A Guide to Creating Healthy Green Growth* (Cambridge, MA: MIT Press, 2021).

8. Wellington Management (August 2021), "Adapting to Climate Change: Investing in the Resiliency Imperative."

9. So long as most debt is held domestically or in the extraordinary privilege of the US, there is an unlimited demand abroad to hold dollars.

10. Guy Standing, *Plunder of the Commons: A Manifesto for Sharing Public Wealth* (London: Pelican, 2019).

11. "The Commons, the State and the Public: A Latin American Perspective," an interview with Daniel Chavez, Transnational Institute (January 10, 2019).

12. Bauwens, Michel, Vasilis Kostakis, and Alex Pazaitis. *Peer to Peer: The Commons Manifesto*. University of Westminster Press, 2019.

13. Barnes, Peter. *Capitalism 3.0: A Guide to Reclaiming the Commons*. Berrett-Kohler, 2014.

14. Meadows, Donella H. "Places to Intervene in a System." *Whole Earth*, 1997.

Chapter 9: A Call to Action

1. G20: Argentina, Australia, Brazil, Canada, China, France, Germany, Great Britain, India, Indonesia, Italy, Japan, Mexico, Russia, Saudi Arabia, South Africa, South Korea, Turkey, the United States, and the countries of the European Union.

Appendix 1: The Earth4All Model

1. J. Randers, *2052: A Global Forecast for the Next Forty Years* (VT: Chelsea Green, 2012).

2. J. Randers et al., *Transformation Is Feasible! How to Achieve the Sustainable Development Goals Within Planetary Boundaries* (Stockholm Resilience Center: Stockholm, 2018).

3. J. Randers et al., "A User-friendly Earth System Model of Low Complexity: The ESCIMO System Dynamics Model of Global Warming Towards 2100," *Earth System Dynamics* 7 (2016): 831–50.

4. D. Collste et al., "Human Well-being in the Anthropocene: Limits to Growth," *Global Sustainability* 4 (2021): 1–17.

5. Narasimha D. Rao, Bas J. van Ruijven, Keywan Riahi, and Valentina Bosetti, "Improving Poverty and Inequality Modelling in Climate Research," *Nature Climate Change* 7, no. 12 (2017): 857–62.

6. Michael Harfoot et al., "Integrated Assessment Models for Ecologists: The Present and the Future," *Global Ecology and Biogeography* 23, no. 2 (2014): 124–43.

7. Mariana Mazzucato, "Financing the Green New Deal," *Nature Sustainability* (2021): 1–2.

8. Stefano Battiston, Irene Monasterolo, Keywan Riahi, and Bas J. van Ruijven, "Accounting for Finance Is Key for Climate Mitigation Pathways," *Science* 372, no. 6545 (2021): 918–20.

9. Tommaso Ciarli and Maria Savona, "Modelling the Evolution of Economic Structure and Climate Change: A Review," *Ecological Economics* 158 (2019): 51–64.

10. Victor Court and Florent McIsaac, "A Representation of the World Population Dynamics for Integrated Assessment Models," *Environmental Modeling & Assessment* 25, no. 5 (2020): 611–32.

11. Efrat Eizenberg and Yosef Jabareen, "Social Sustainability: A New Conceptual Framework," *Sustainability* 9, no. 1 (2017): 68.

Index

About the Authors

Sandrine Dixson-Declève is co-president of the Club of Rome and has over 30 years of leadership in climate change, sustainability, innovation, and energy. GreenBiz named her one of the 30 most influential women driving change in the low-carbon economy. She is a policy advisor, facilitator, TED speaker, teacher, and author of *Quel Monde Pour Demain?*

Owen Gaffney is a changemaker, strategist, writer, filmmaker, and global sustainability analyst at Stockholm Resilience Centre and Potsdam Institute for Climate Impact Research. He is co-founder of the Exponential Roadmap Initiative and has written, produced and advised on multimedia and documentaries for the BBC, Netflix, TED, WWF and the World Economic Forum.

Jayati Ghosh is an internationally recognized development economist and professor at the University of Massachusetts. She has authored and/or edited 19 books and nearly 200 scholarly articles, received several national and international prizes, and is member of an array of international commissions. She writes regularly for a variety of media.

Jorgen Randers is professor emeritus of climate strategy at the BI Norwegian Business School. A global leader on the intersections of economy, the environment, and human wellbeing, he coauthored *The Limits to Growth* in 1972 and the 30-year update. He is the author of *2052* and coauthor of *Reinventing Prosperity* and *Transformation Is Feasible!*

Johan Rockström is director of Potsdam Institute for Climate Impact Research. He led the team of scientists that proposed the influential planetary boundaries framework. Rockström's TED talks have received over five million views. He is the subject of the Netflix documentary *Breaking Boundaries*, narrated by David Attenborough.

Per Espen Stoknes directs the Center for Sustainability and Energy at the BI Norwegian Business School. He is a TED speaker, and was an MP in the Norwegian parliament (2017–2018), has co-founded clean-energy companies, and is the author of several books, including *Tomorrow's Economy* and *What We Think About When We Try Not to Think About Global Warming.*

ABOUT NEW SOCIETY PUBLISHERS

New Society Publishers is an activist, solutions-oriented publisher focused on publishing books to build a more just and sustainable future. Our books offer tips, tools, and insights from leading experts in a wide range of areas.

We're proud to hold to the highest environmental and social standards of any publisher in North America. When you buy New Society books, you are part of the solution!

At New Society Publishers, we care deeply about *what* we publish—but also about *how* we do business.

- Most of our books are printed on 100% **post-consumer recycled paper**, processed chlorine-free, with low-VOC vegetable-based inks (since 2002).

- Our corporate structure is an innovative employee shareholder agreement, so we're one-third employee-owned (since 2015)

- We've created a Statement of Ethics (2021). The intent of this Statement is to act as a framework to guide our actions and facilitate feedback for continuous improvement of our work

- We're carbon-neutral (since 2006)

- We're certified as a B Corporation (since 2016)

- We're Signatories to the UN's Sustainable Development Goals (SDG) Publishers Compact (2020–2030, the Decade of Action)

To download our full catalog, sign up for our quarterly newsletter, and to learn more about New Society Publishers, please visit newsociety.com

CPSIA information can be obtained
at www.ICGtesting.com
Printed in the USA
BVHW051659310123
657543BV00025B/544